BLUEPRINT READING
====for the====
BUILDING TRADES

BLUEPRINT
═══ READING ═══
for the
BUILDING TRADES

John E. Traister

Craftsman Book Company
6058 Corte Del Cedro, Carlsbad, CA 92009

Library of Congress Cataloging in Publication Data

Traister, John E.
   Blueprint reading for the building trades.

   Reprint. Originally published; Reston, Va. ;
Reston Pub. Co., c1981.
   Includes index.
   1. Building--Details--Drawings.   2. Blueprints.
I. Title.
TH431.T72   1985   692'.1   85-19455
ISBN 0-934041-05-9

Contents

Preface

The accelerated growth of the building construction industry in the United States and other advanced nations has brought with it the need for greater numbers of practical craftsmen, technicians, and design engineers, all of who need to know how to interpret building construction drawings: namely, working drawings and written specifications. This book is designed to provide the reader with the instructional material necessary to gain a proper knowledge of blueprint reading as it is related to the building construction industry.

The chapters are arranged in a structural sequence intended to satisfy the immediate and fundamental needs of vocational students, apprentices in the buildings trades, beginning students in engineering schools, as well as workers already engaged in the building construction industry who wish to improve their working knowledge of blueprint reading.

Upon completion of this book, the reader should be able to correctly interpret all types of trade drawings in order to plan the installation of the reader's chosen trade or profession.

John E. Traister

1 Building Construction Documents

Objectives

The fundamental objective of this chapter is to give the reader an overall picture of the building construction industry and its relationship to construction documents. Familiarity with this relationship is considered necessary to give the reader a proper background for approaching the subject of blueprint reading.

The Construction of a Building

In all large construction projects, and in most of the smaller ones also, an architect is commissioned to prepare complete working drawings and specifications for the project. These drawings usually include:

1. A plot plan indicating the location of the building on the property, as shown in Fig. 1-1.

2. Floor plans showing the walls and partitions for each floor or level, as shown in Fig. 1-2.

3. Elevations of all exterior faces of the building, as shown in Fig. 1-3.

4. A number of vertical cross sections to indicate clearly the various floor levels and details of the footings, foundation, walls, floors, ceilings, and roof construction.

5. Large-scale detail drawings showing such details of construction as may be required.

For projects of any consequence, the architect usually hires consulting engineers to prepare structural, electrical, plumbing, heating, ventilating, and air-conditioning drawings. A brief description of such drawings follows.

Structural Drawings: A typical structural drawing is shown in Fig. 1-4. Such drawings are most often prepared by structural engineers on

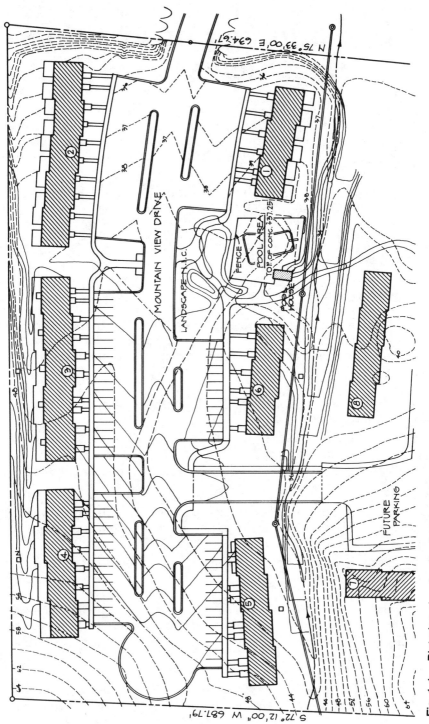

Fig. 1-1. Plot plan showing the location of buildings on the property and all outside utilities.

2

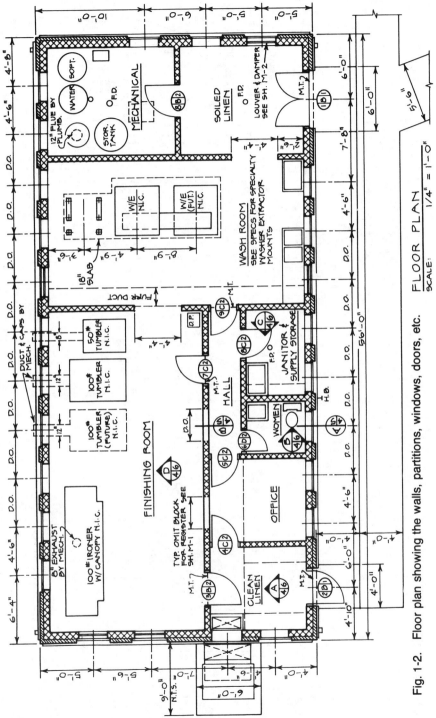

Fig. 1-2. Floor plan showing the walls, partitions, windows, doors, etc.

3

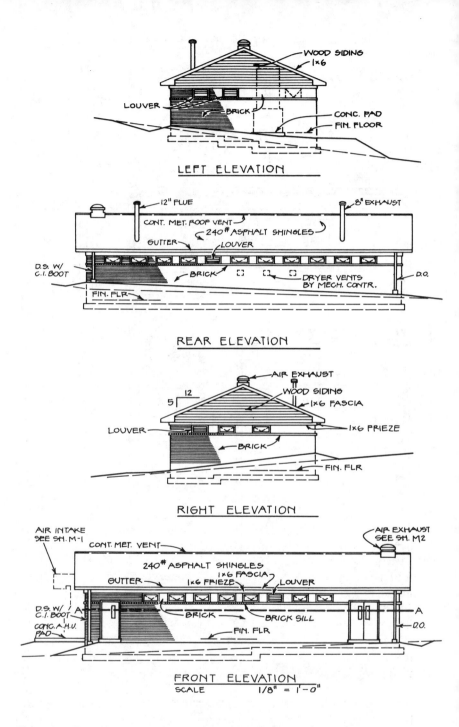

Fig. 1-3. Four elevation drawings showing all exterior faces of a building.

4

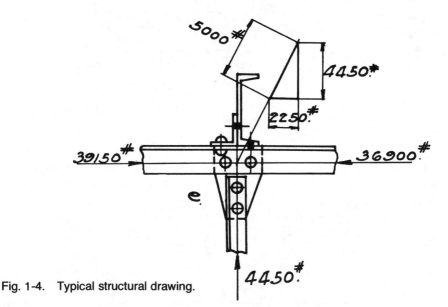

Fig. 1-4. Typical structural drawing.

the basis of proper allowances for all vertical loads and lateral stresses and are included with the architectural drawings for all long-span, wood-truss construction and all reinforced concrete and structural steel construction.

Electrical Drawings: The electrical drawings for a building project generally cover the complete electrical design of the electrical system for lighting, power, alarm and communication systems, special electrical systems, and related electrical equipment. These drawings sometimes include a plot plan or site plan showing the location of the building on the property and the interconnecting electrical systems; floor plans showing the location of all outlets, lighting fixtures, panelboards, and other components and equipment; power-riser diagrams; a symbol list; schematic diagrams; and large-scale details where necessary.

Mechanical Drawings: Mechanical drawings cover the installation of the plumbing, heating, ventilating, and air-conditioning systems within a building and on the premises. They cover the complete design and layout of these systems and show floor-plan layouts, cross sections of the building, and necessary detailed drawings. Control wiring for various heating and air-conditioning controls may also be included on the mechanical drawings.

A typical electrical drawing is shown in Fig. 1-5, a plumbing drawing is shown in Fig. 1-6, and a typical air-conditioning drawing is illustrated in Fig. 1-7. The reader is not expected to understand every detail of these drawings at this time. However, the reader should re-

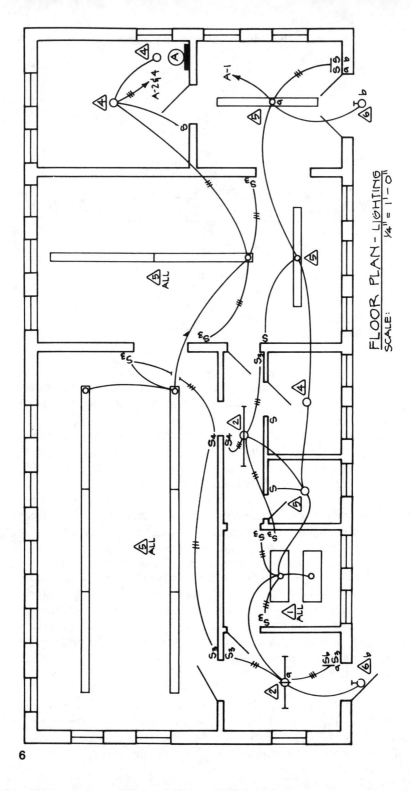

FLOOR PLAN - LIGHTING
SCALE: ¼" = 1'-0"

Fig. 1-5.   Typical electrical drawing showing lighting outlets and related wiring.

6

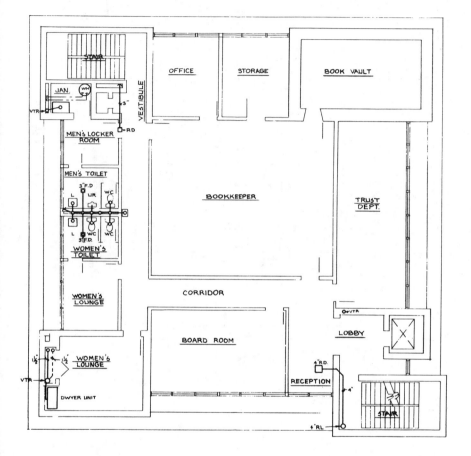

Fig. 1-6.   Floor plan showing the plumbing layout for a building.

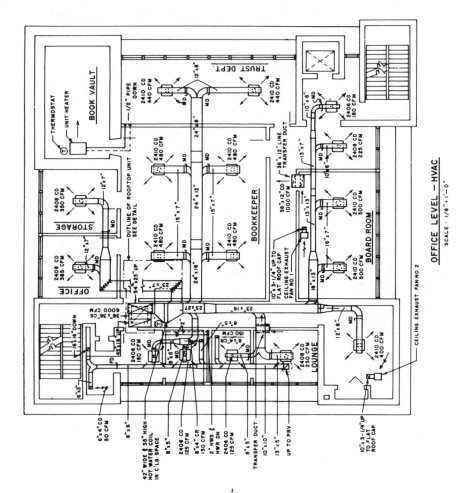

Fig. 1-7. Floor plan showing the air conditioning system of a building.

view these drawings, noting every detail, to get an overall picture of the various types of construction drawings in use. Then, subsequent chapters will tend to clear up any hazy areas as the reader progresses.

Construction Specifications

Construction specifications for a building construction project are the written description of the work and duties required of the owner, the architect, and the engineers. Together with the working drawings, these specifications form a basis of the contract requirements for the construction of the building or project.

A sample page from an electrical specification is shown in Fig. 1-8. Further details on written specifications along with many other examples may be found in Chapter 9, Construction Specifications.

When all of these documents are completed, the architect will often represent the owner in soliciting quotations from general contractors and advise the owner as to the proper award to make. The architect will also usually represent the owner during construction of the building, inspecting the work to ascertain that it is being performed in accordance with the requirements of the drawings and specifications.

The specialty work, such as electrical and mechanical systems, is also under the architect's responsibility. However, the architect's consulting engineers will help in soliciting bids from the electrical and mechanical contractors. The engineer also inspects his portion of the work to assure the architect and owner that this portion is carried out according to the working drawings and specifications.

The architects and engineers usually approve shop drawings (Fig. 1-9), check and approve progress payments, and perform many other duties during the construction of the project. In other words, they see the project through from beginning to end.

Types of Construction Drawings

Those involved in the building construction industry will encounter many types of drawings during the construction of even a single project. Therefore, a brief sampling of the various types likely to be encountered is in order. A more detailed study will be covered in the chapters to follow.

Pictorial drawings: These drawings are those in which an object is drawn in one view only, that is, three dimensional effects are simulated on the flat plane of drawing paper by drawing several faces of an object in a single view. This type of drawing is used to convey information to those not trained in blueprint reading, or to supplement the conventional orthographic drawings in the more complex systems.

DIVISION 16 - ELECTRICAL

SECTION 16A - GENERAL PROVISIONS

1. Portions of the sections of the Documents designated by the letters "A", "B" & "C" and "DIVISION ONE - GENERAL REQUIREMENTS" apply to this Division.

2. Consult Index to be certain that set of Documents and Specifications is complete. Report omissions or discrepancies to the Architect.

3. SCOPE OF THE WORK:

a. The scope of the work consists of the furnishing and installing of complete electrical systems - exterior and interior - including miscellaneous systems. The Electrical Contractor shall provide all supervision, labor, materials, equipment, machinery, and any and all other items necessary to complete the systems. The Electrical Contractor shall note that all items of equipment are specified in the singular; however, the Contractor shall provide and install the number of items of equipment as indicated on the drawings and as required for complete systems.

b. It is the intention of the Specifications and Drawings to call for finished work, tested, and ready for operation.

c. Any apparatus, appliance, material or work not shown on drawings but mentioned in the specifications, or vice versa, or any incidental accessories necessary to make the work complete and perfect in all respects and ready for operation, even if not particularly specified, shall be furnished, delivered and installed by the Contractor without additional expense to the Owner.

d. Minor details not usually shown or specified, but necessary for proper installation and operation, shall be included in the Contractor's estimate, the same as if herein specified or shown.

Fig. 1-8.   Sample page from a set of electrical specifications.

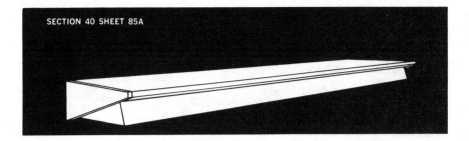

SECTION 40 SHEET 85A

SPECIFICATIONS:

APPLICATION: Wall-mounted fixture providing downward light as well as shelf area.

CONSTRUCTION: Base structure is a triangular solid of die-formed steel with no exposed mechanical fasteners. Solid top provides shelf area. Complete units in 2-, 3- and 4-foot lengths.

FINISH: Hot-bonded baked white enamel. Units treated with five-stage coating of zinc-phosphate. Unpainted parts protected by bright-dip zinc plate.

WIRING: Internal thermally protected HPF single lamp ballast. Ballast for 40-watt rapid-start lamps to be CBM certified. Ballast for 30-watt to be rapid-start; 20-watt trigger-start. Toggle switch wired in but not mounted.

INSTALLATION: Knock-outs in back for mounting direct to wall.

Catalog Number	One-Lamp Complete Units	Overall Length	Approximate Shipping Weight
8121W†	One-Lamp Baked White Unit for 2-ft. Trigger-Start	25″	8 lbs.
8221W†	Two-Lamp Baked White Unit for 2-ft. Trigger-Start	25″	8 lbs.
8131W†	One-Lamp Baked White Unit for 3-ft. Rapid-Start.	37″	14 lbs.
8141W†	One-Lamp Baked White Unit for 4-ft. Rapid-Start.	49″	21 lbs.
BE81	Pair Adjustable Sliding Book Ends*		

*To be installed during original installation.
†For prime white coat for on-the-job painting add suffix "P" in place of "W".

AVERAGE FOOTCANDLES ON DESK

	MOUNTING HEIGHT	
	2'0	2'6"
2-FT. (1 LAMP)	60	45
2-FT. (2 LAMP)	100	80
3-FT. (1 LAMP)	75	65
4-FT. (1 LAMP)	80	75

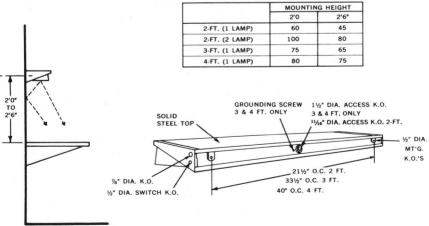

Fig. 1-9.   A typical shop drawing received for approval by architect and engineers.

The types of pictorial drawings most often found in the building construction fields include:

1. isometric
2. oblique
3. perspective

The main disadvantages of pictorial drawings are that intricate parts cannot be pictured clearly and that they are difficult to dimension.

A view projected onto a vertical plane in which all of the edges are foreshortened equally is called an *isometric projection*. Fig. 1-10 shows an isometric drawing of a cube. In this view, the edges are 120 degrees apart and are called the isometric axes, while the three surfaces shown are called the isometric planes. The lines parallel to the isometric axes are called the isometric lines.

This type of pictorial drawing is usually preferred over the other two types mentioned for building construction drawings, because it is possible to draw isometric lines to scale in the same manner as floor plans or other multiview plans drawn in orthographic views. An example of a plumbing pipe riser diagram, utilizing isometric techniques, is shown in Fig. 1-11.

The *oblique drawing* is similar to the isometric drawing in that one face of the object is drawn in its true shape and the other visible faces are shown by parallel lines drawn at the same angle (usually 45-30 degrees) with the horizontal. However, unlike an isometric drawing,

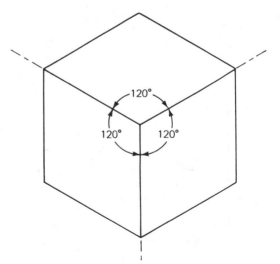

Fig. 1-10.   Isometric drawing of a cube.

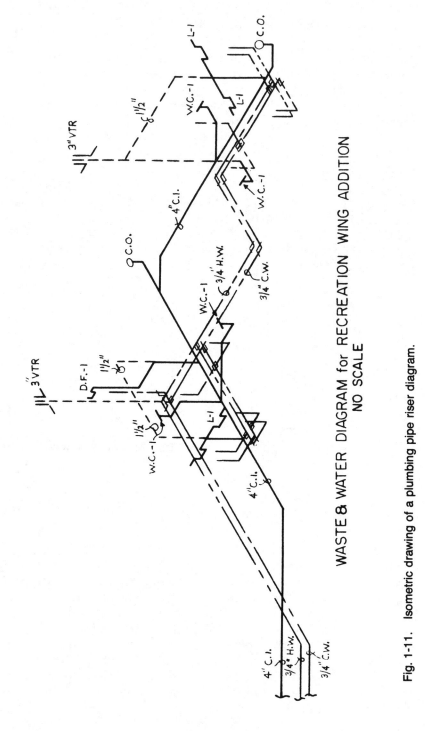

WASTE & WATER DIAGRAM for RECREATION WING ADDITION
NO SCALE

Fig. 1-11.   Isometric drawing of a plumbing pipe riser diagram.

13

the lines drawn at a 30-degree angle are shortened to preserve the appearance of the object and are therefore not drawn to scale. The drawing in Fig. 1-12 is an oblique drawing of a cube.

Sometimes it is desired to show an object in an exact pictorial representation—as it appears to the eye. A drawing of this type is called a *perspective drawing.* A typical use of such a drawing is for showing a representation of a building as it will look when it is completed. Such a drawing appears in Fig. 1-13.

Orthographic-projection drawings: This type represents the physical arrangement and views of specific objects. Those working in the building construction trades will encounter this type of drawing more than any other. In general, orthographic-projection drawings give all plan views, elevation views, dimensions, and other details necessary to construct the project or object.

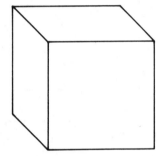

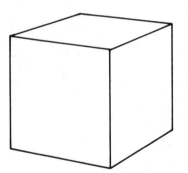

Fig. 1-12.   An oblique drawing of a cube.

Fig. 1-13.   A perspective drawing of a cube.

To illustrate the practicability of the orthographic drawing, look at the pictorial drawing in Fig. 1-14. While this view clearly suggests the form of a block, it does not show the actual shape of the surfaces, nor does it show the dimensions of the object so that it may be constructed.

An orthographic projection of this same block appears in Fig. 1-15. One of the drawings in this figure shows the block as though the observer were looking straight at the front; one as though the observer were looking straight at the left side; one as though the observer were looking straight at the right side; and one as though the observer were looking at the rear of the block. The remaining view is as if the observer were looking straight down on the block. These views, combined with the dimensions, will allow the object to be constructed properly from materials such as metal, wood, plastic, etc. This material may be

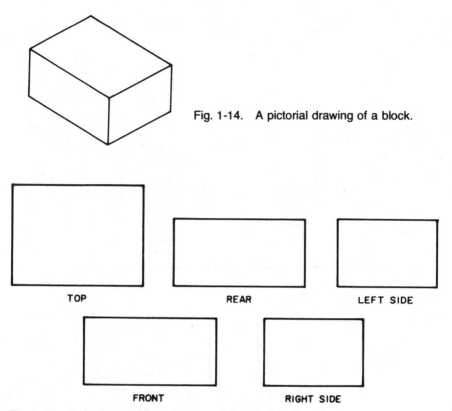

Fig. 1-14.   A pictorial drawing of a block.

TOP

REAR

LEFT SIDE

FRONT

RIGHT SIDE

Fig. 1-15.   An orthographic projection of the block in Fig. 1-14.

specified in written specifications, a schedule, or by merely adding a note to the drawing.

Diagrams: Diagrams are drawings that are intended to show, in diagrammatic form, components and their related connections. Such drawings are seldom drawn to scale, and show only the working association of the different components. In diagram drawings, symbols are used extensively to represent the various pieces of electrical components, and lines are used to connect these symbols—indicating the size, type, and number of components or pieces of equipment.

A typical flow diagram is shown in Fig. 1-16 while a typical electrical schematic diagram is shown in Fig. 1-17.

Need of Construction Documents

By now it should be obvious that construction working drawings consist of lines, symbols, dimensions, and notations to accurately convey an architect's or engineer's design to workmen who construct the

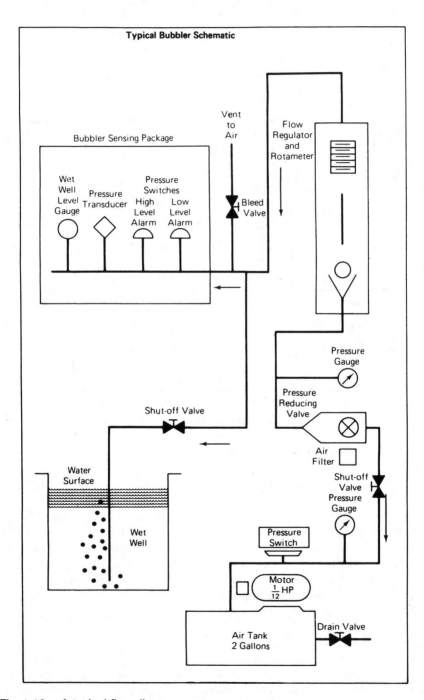

Fig. 1-16.   A typical flow diagram.

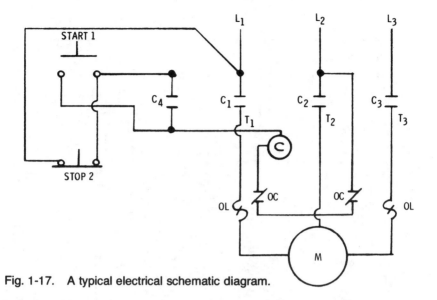

Fig. 1-17. A typical electrical schematic diagram.

projects. The workmen must be able to take the construction documents (drawings and specifications), and without further instructions, install or construct the various systems as the architect or engineer intended it to be accomplished. A working drawing, therefore, is an abbreviated language for conveying a large amount of exact, detailed information, which would otherwise take many pages of manuscript or hours of verbal instruction to convey.

In every branch of the building construction industry, there is often occasion to read construction drawings. Electricians, for example, who are responsible for installing the electrical system in a new building, usually consult an electrical drawing to locate the various outlets, the routing of circuits, the location and size of panelboards, switches, and the like. Laborers must consult the site plan for the routing of trenches for the building's utilities. Air-conditioning personnel must consult drawings for fabricating the ductwork, location of equipment, etc. The list is almost endless.

Practical Application of Working Drawings

You previously reviewed the various types of drawings most often encountered in the building construction industry, but many of the examples shown—like the cubes and blocks—are hardly the typical drawings the worker will encounter on an actual project. The examples to follow are intended to show how these previous examples relate to actual applications in the building construction field.

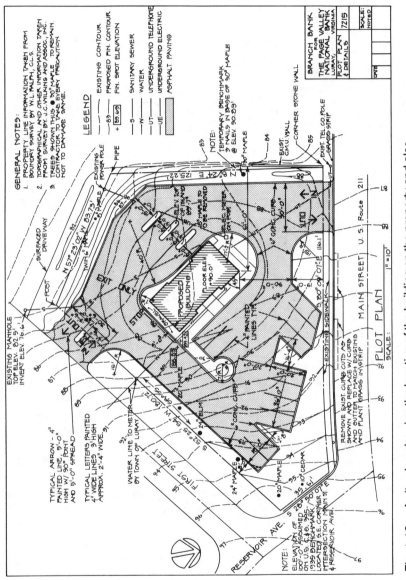

Fig. 1-18. A site plan showing the location of the building on the property and also all outside utilities, roads, boundary lines, contour lines, and the like.

18

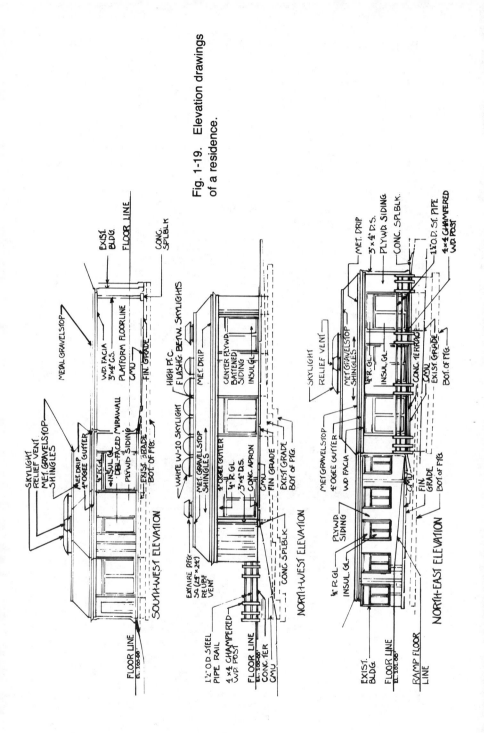

Fig. 1-19. Elevation drawings of a residence.

19

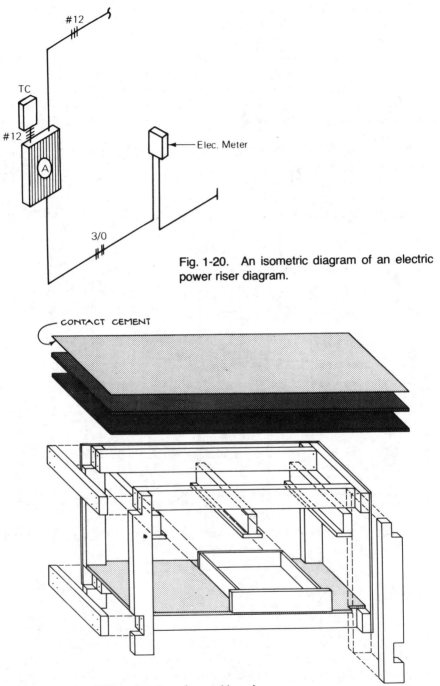

Fig. 1-20. An isometric diagram of an electric power riser diagram.

Fig. 1-21. An oblique drawing of a workbench.

The orthographic drawing in Fig. 1-18 is a site plan showing the location of buildings on the property and all outside utilities, roads, boundary lines, contour lines, and the like. Note that all of this information is conveyed by means of lines, symbols, and notes. An experienced person can look at this one drawing and tell how much excavation is required to construct the roads and building foundation; the location of all sewage manholes and the length and routing of pipe between them; the distance between the buildings and the property lines; and a host of other details.

The example of an orthographic projection in Fig. 1-15 may be compared to elevation drawings of a residence shown in Fig. 1-19. An isometric diagram of an electric power riser diagram is shown in Fig. 1-20, while an oblique drawing of a workbench is shown in Fig. 1-21. A schematic diagram of heating and cooling wiring is shown in Fig. 1-22.

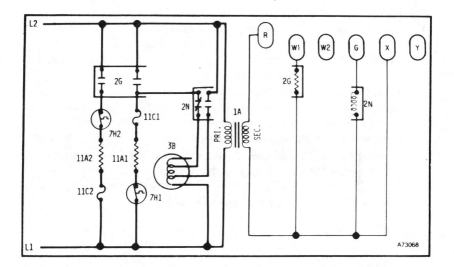

Fig. 1-22. A schematic diagram of a control circuit for a heating and cooling system.

Summary

A building construction drawing is really an abbreviated language for conveying a large amount of exact, detailed information, which would otherwise take many pages of manuscript or hours of verbal instruction to convey. Drawings themselves consist of lines, symbols, and notations to convey the information.

A set of construction documents will usually consist of a plot plan, floor plans of each floor or level, sectional drawings, various large-scale details, elevation drawings, and drawings for the specialty trades like structural, electrical, and mechanical.

Questions

1. A pictorial drawing of an object as it actually appears to the eye is called a _____ drawing.
2. Drawings that are intended to show, in diagrammatic form, components and their related connections are called _____.
3. A plot plan normally shows the location of the _____ in relation to the property lines.
4. On large projects, architects will usually hire a _____ _____ to design and prepare drawings for the electrical and mechanical systems.
5. Proper allowances for vertical loads and lateral stresses are normally shown on _____ drawings.
6. Construction specifications are a _____ description of the work and duties required by the contractor.
7. When a project is ready for bids, the _____ will often represent the owner in soliciting quotations from contractors.
8. A view projected onto a vertical plane in which all of the edges are foreshortened equally is called a(n) _____ drawing.
9. The type of pictorial drawing usually preferred for construction drawings is the _____ drawing.
10. A(n) _____ drawing is one that represents the physical arrangement and views of specific objects.

Answers to Questions

1. perspective
2. diagrams
3. building(s)
4. consulting engineer
5. structural
6. written
7. architect
8. isometric
9. isometric
10. orthographic-projection

2 Layout of Construction Documents

Objectives

To show the reader how construction documents are prepared.

When an architect is commissioned to prepare complete working drawings and specifications for a building, a conference is held with the owners so that the architectural firm can determine what type of building is wanted, what functions will be performed in the building, the budget allotted for the project, and other pertinent data. During this preliminary conference, the architect will also make recommendations to the owners and decide if such a project is feasible.

Once the initial information is obtained, the architect and his staff (including consulting engineers) will research the project thoroughly and sketch several preliminary drawings for the owner's review. Other conferences between the owners and architect will be held—using the preliminary drawings as a guide—to tie the design down to more exact details; that is, certain areas will be omitted, others will be added while still other areas will be modified. Depending upon the size and complexity of the project, approval of the preliminary drawings may take as little as one day to as long as several months or even years.

After a definite agreement is reached between the owners and the architect, the architectural firm will begin preparing the working drawings and written specifications.

Analyzing a Set of Working Drawings

The most practical way to learn how construction documents are prepared is to analyze an existing drawing prepared by an architectural firm.

The first page of a set of blueprints is called the "title sheet" (see Fig. 2-1). Formats will vary from firm to firm, but in general, this first sheet contains the name and location of the building project, the name of the architectural firm, the name of the consulting engineering firm, an index to the drawings, and usually, the legend or key for materials used in the drawing sheets to follow.

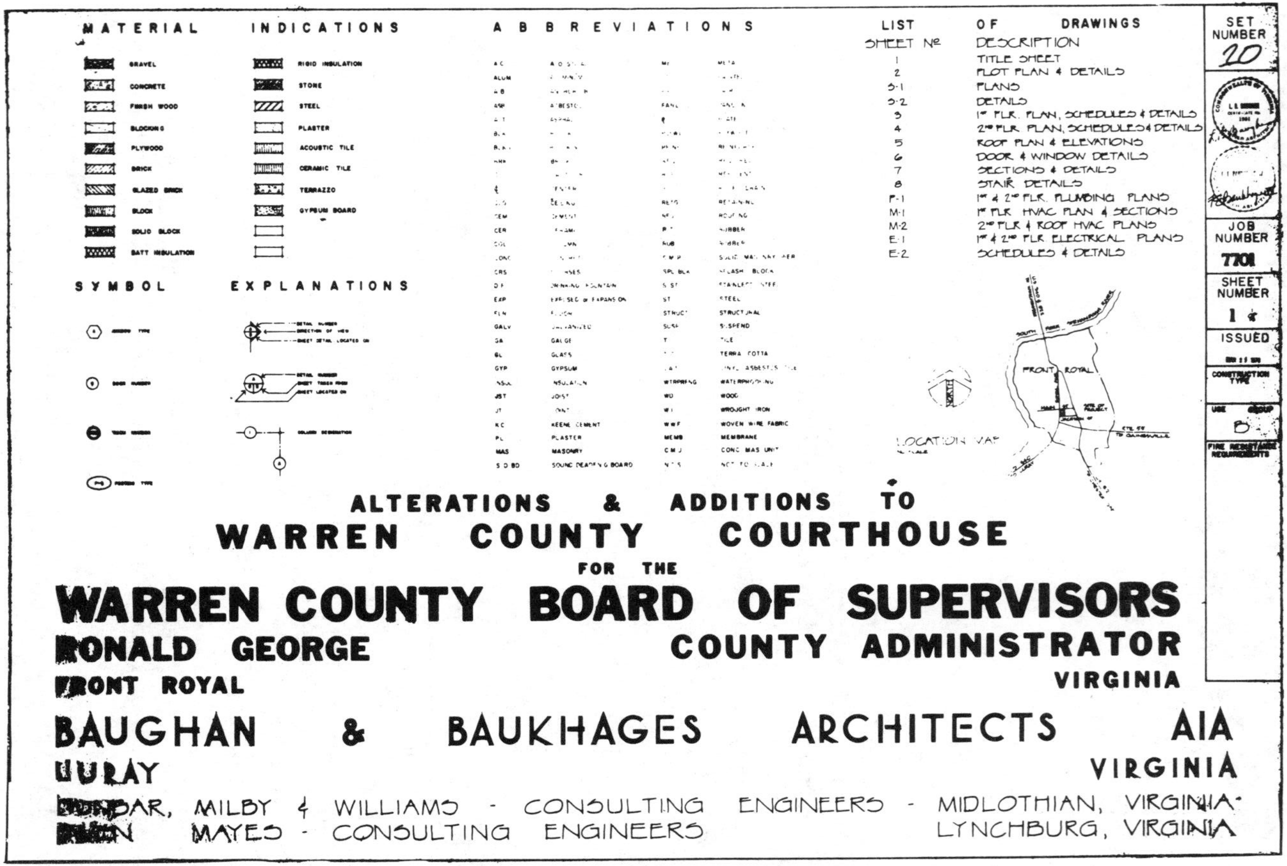

MATERIAL INDICATIONS

GRAVEL — RIGID INSULATION
CONCRETE — STONE
FIBER WOOD — STEEL
BLOCKING — PLASTER
PLYWOOD — ACOUSTIC TILE
BRICK — CERAMIC TILE
GLAZED BRICK — TERRAZZO
BLOCK — GYPSUM BOARD
SOLID BLOCK
BATT INSULATION

SYMBOL EXPLANATIONS

A B B R E V I A T I O N S

LIST OF DRAWINGS

SHEET Nº	DESCRIPTION
1	TITLE SHEET
2	PLOT PLAN & DETAILS
S-1	PLAN
S-2	DETAILS
3	1ˢᵗ FLR. PLAN, SCHEDULES & DETAILS
4	2ᴺᴰ FLR. PLAN, SCHEDULES & DETAILS
5	ROOF PLAN & ELEVATIONS
6	DOOR & WINDOW DETAILS
7	SECTIONS & DETAILS
8	STAIR DETAILS
P-1	1ˢᵗ & 2ᴺᴰ FLR. PLUMBING PLANS
M-1	1ˢᵗ FLR. HVAC PLAN & SECTION
M-2	2ᴺᴰ FLR. & ROOF HVAC PLANS
E-1	1ˢᵗ & 2ᴺᴰ FLR. ELECTRICAL PLANS
E-2	SCHEDULES & DETAILS

LOCATION MAP

ALTERATIONS & ADDITIONS TO
WARREN COUNTY COURTHOUSE
FOR THE
WARREN COUNTY BOARD OF SUPERVISORS
COUNTY ADMINISTRATOR
RONALD GEORGE
FRONT ROYAL VIRGINIA VIRGINIA

BAUGHAN & BAUKHAGES ARCHITECTS AIA
GRAY LYNCHBURG, VIRGINIA

DUNBAR, MILBY & WILLIAMS - CONSULTING ENGINEERS - MIDLOTHIAN, VIRGINIA
REID & MAYES - CONSULTING ENGINEERS LYNCHBURG, VIRGINIA

SET NUMBER 20
JOB NUMBER 7701
SHEET NUMBER 1 of
ISSUED

Fig. 2-1. The first page of a set of construction drawings and is normally called "The Title Sheet."

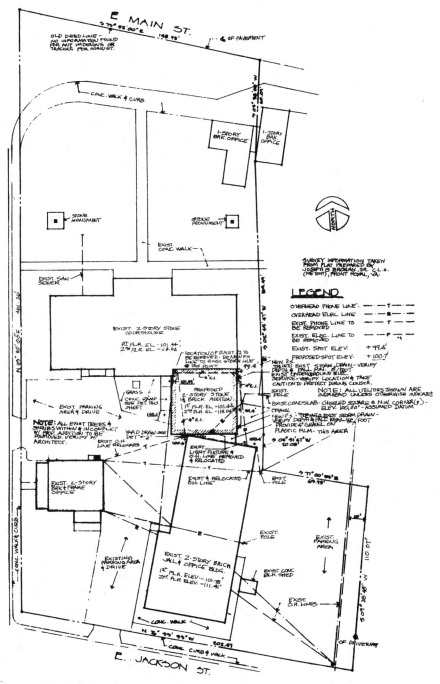

Fig. 2-2. Typical plot plan.

The second page in the set usually includes a drawing of the plot on which the building or buildings are to be constructed. A breakdown of soil borings taken to examine the bearing qualities of the earth may also be included.

A drawing of a typical plot plan is shown in Fig. 2-2. This plan is drawn to scale with pertinent dimensions and indicates the location of the building on the site as well as its walks and drives. Natural ground slope of the side is indicated by grade lines. The drawing also shows what the "finish grade" of the site is to be when the building is erected and certain earth moving has been accomplished.

The elevation figures on the plot plan in Fig. 2-2 are based on a "bench mark," which is a permanent object in the area with a pre-determined elevation in feet above sea level. A typical bench mark is shown in Fig. 2-3. In all other drawings in the set that follow, the first-floor level of the building is given a base elevation of 100 feet for the purpose of easier dimensioning. Then other elevations are figured from that. The plot plan also contains indications of existing buildings, streets, trees, and permanent landmarks.

On smaller projects, utility services such as water, electricity, telephone, cable TV, and gas and sewer lines are also indicated on the plot plan. On larger projects, a separate sheet titled "Utility Plot Plan" will be included to facilitate the reading of the drawings.

Fig. 2-3. Typical bench mark.

Drawing sheets immediately following the plot plan will normally contain the building floor plans, examples of which appear in Fig. 2-4. Floor plans, more so than any other type of drawing, show how the building is laid out, and is the most readable to the nonprofessional.

A floor plan represents a cut horizontally through a building at approximately eye level and appears on the sheet as though the top half of the cut is removed and the viewer is looking down on the building from above. A separate drawing is made for each floor or level (including

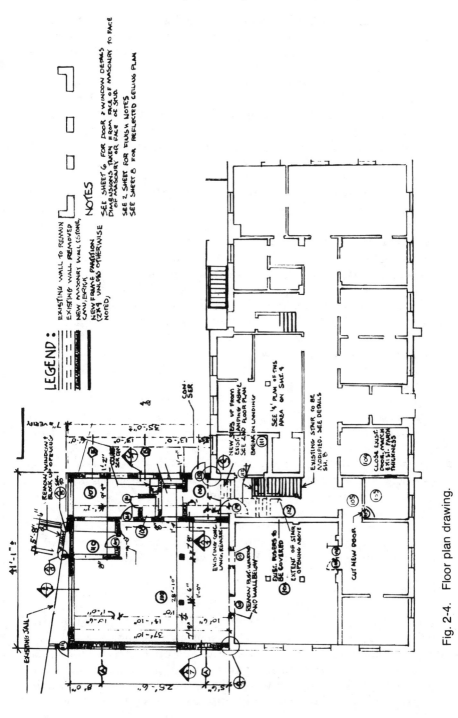

Fig. 2-4.   Floor plan drawing.

basement). Shown in the floor plans are the location and arrangement of all walls, partitions, doors, windows, and stairways, with indication of dimensions and materials used. Floor plans also contain many references to elevation, section, and other plan drawings on other sheets and could be called the key sheet of a set of working drawings.

Following the floor plans are drawing sheets showing the building elevations. Elevations are head-on, vertical views of a building or wall area in a single plane. Therefore, the front elevation of a building is as if the viewer looked straight at the building from the front; the right side elevation is as if the viewer looked straight at the right side of the building, and so forth.

Elevations also show the materials with which the walls are constructed, that is, glass, concrete block, etc. Symbols (Chapter 3) are used to indicate most of the finishes although notes are sometimes employed. Figure 2-5 shows examples of elevation drawings.

Drawing sheets showing various cross sections appear in Fig. 2-6. These cross sections appear as if the area of the building in question is sliced open to reveal construction details that can't be shown by drawings of elevations or plans. Both small- and large-scale sections are necessary throughout a set of blueprints to enable workers to construct the building. A thorough study of cross sections and other views is a great aid in learning how a building is constructed, since great detail is shown.

The parts of the working drawing discussed thus far show most of the construction details. However, certain construction conditions cannot be shown adequately in the scale in which most drawings are made. Therefore, larger-scaled drawings with greater detail must sometimes be made of some areas to ensure that the workmen will understand what is to be done. These are termed special, or large-scale, detail drawings and are found throughout a set of working drawings to better explain construction and design features.

A thorough study of construction details, such as those found in Chapter 5, is helpful in learning to read blueprints, because such drawings show construction conditions clearly and add to the ability to visualize the finished product from the working drawing.

Structural Drawings

Following the architectural drawings—usually designated by sheet numbers A-1, A-2, etc.—are structural drawings showing details of footings, foundation, structural framing, and other structural details.

Footings are concrete "feet," placed in the ground and sometimes reinforced with steel bars on which the foundation and the subsequent building load is placed. Soil conditions, together with the weight of the

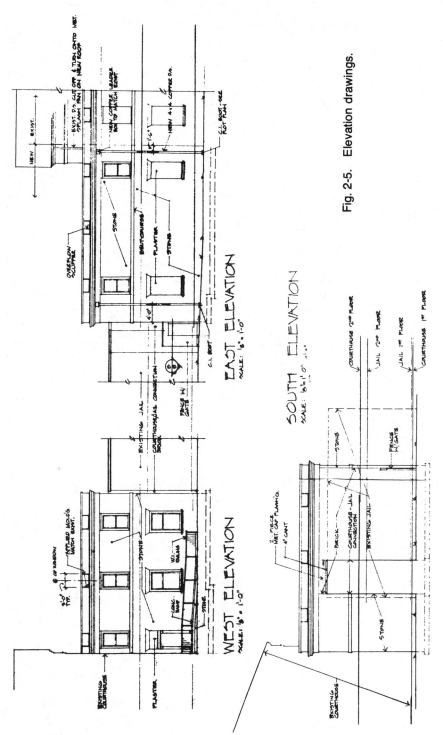

Fig. 2-5. Elevation drawings.

29

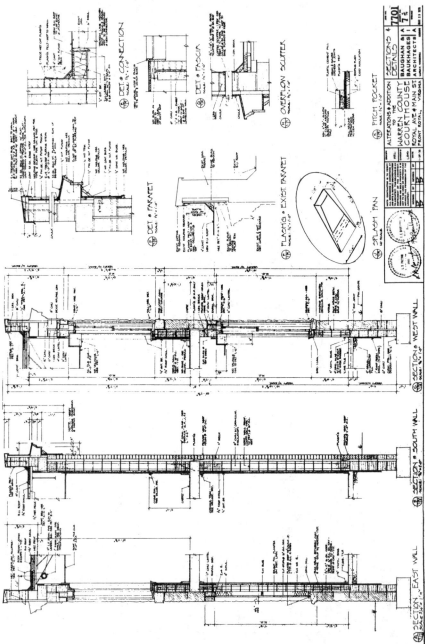

Fig. 2-6. Cross sections of a building.

30

building, determine the size, design, and number of footings. Building weight is measured in terms of "dead" and "live" loads. Dead load is the stationary weight of the building itself and the permanently fixed equipment. Live load is made up of movable equipment in the building and humans who use it.

Foundation walls serve as a base on which the building is built, and carry the load of the building to the footings and earth below. These walls must be strong enough to resist the side pressure of the earth. Steel reinforcing bars are used in foundation walls, which are deep into the ground to offset the extra heavy earth pressure. Another function of the walls is to keep moisture out of the underground parts of the building. Usually a waterproofing compound is applied to the exterior surface of foundation walls.

Figures 2-7, 2-8, and 2-9 show examples of structural drawings used on actual installations. Note the details, schedules, and other pertinent data included.

Mechanical Drawings

Mechanical drawings are usually included in all sets of working drawings for projects of any consequence. These drawings include details for the plumbing, heating, ventilating, air-conditioning, and temperature control systems.

Most architects hire consulting engineers who specialize in this field to design and prepare working drawings and specifications for this portion of the project. In doing so the design must be coordinated very carefully with the architect as well as the electrical designer to ensure smooth integration of the proper facilities with the building design.

Mechanical drawings are highly diagrammatical and are used to locate pipes, fixtures, ductwork, equipment, and so forth. Detailed descriptions of this and other equipment are found in the written specifications. Fig. 2-10 shows a typical plumbing drawing and Fig. 2-11 shows a heating and cooling system layout.

Electrical Drawings

Electrical drawings are prepared in much the same way as mechanical drawings; that is, architects hire consulting engineers to design the system and then prepare working drawings and written specifications.

Electrical drawings prepared by consulting engineers are unique drawings. Most sets of electrical drawings encompass all of the techniques described in Chapter 1. For example, a complete set of working drawings for an electrical system will usually consist of the following:

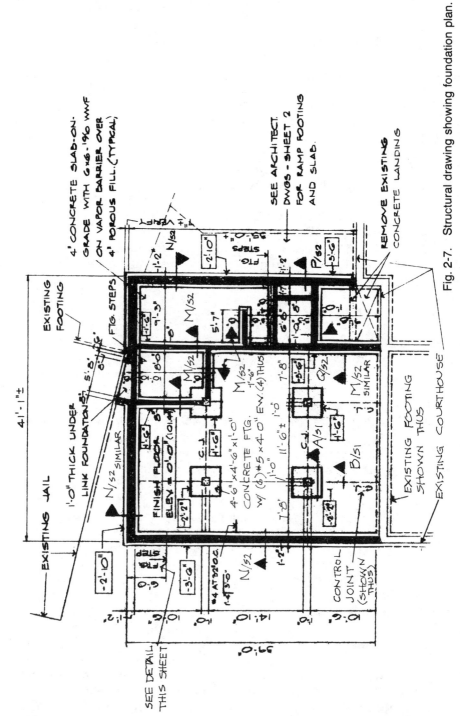

Fig. 2-7. Structural drawing showing foundation plan.

32

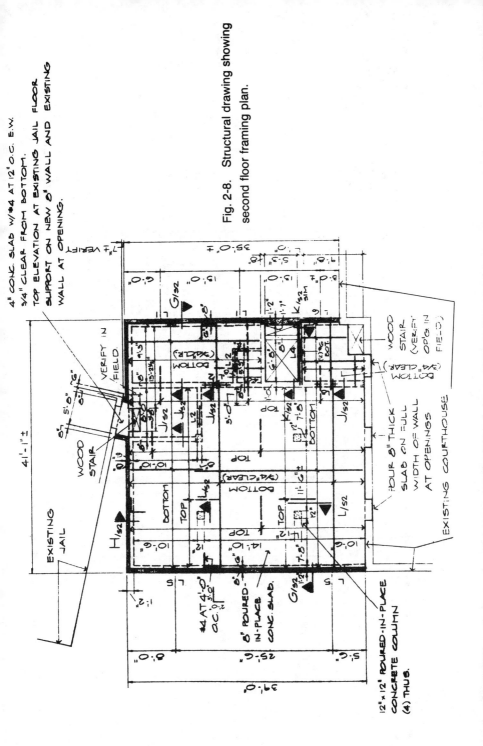

Fig. 2-8. Structural drawing showing second floor framing plan.

33

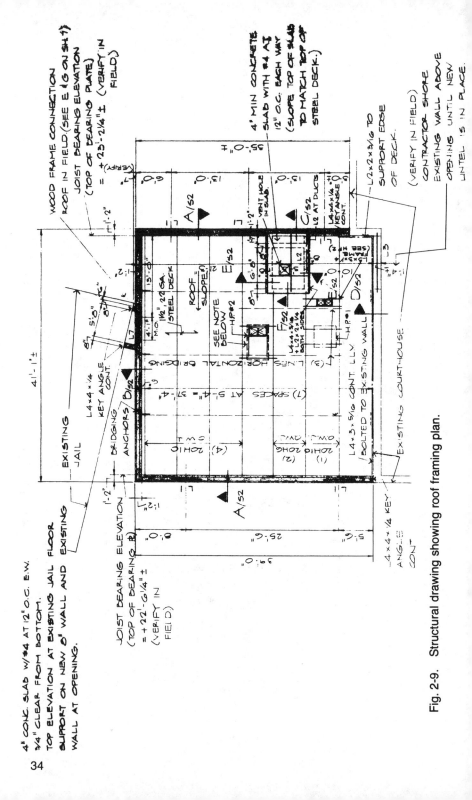

Fig. 2-9. Structural drawing showing roof framing plan.

34

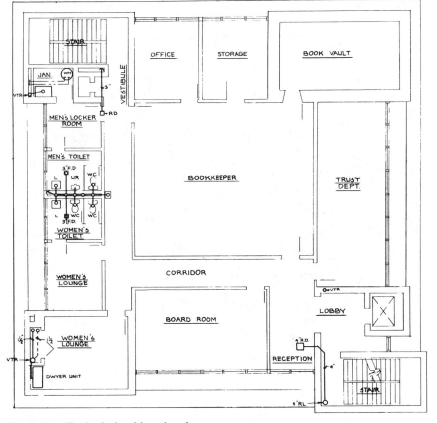

Fig. 2-10. Typical plumbing drawing.

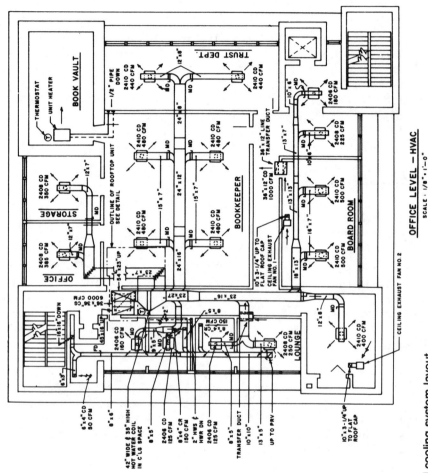

Fig. 2-11. Heating and cooling system layout.

36

1.   A plot plan showing the location of the building on the property and all outside electrical wiring, including the service entrance. This plan is drawn to scale with the exception of the various electrical symbols, which must be enlarged to be readable.

2.   Floor plans showing the walls and partitions for each floor level. The physical locations of all wiring and outlets are shown for lighting, power, signal and communication, special electrical systems, and related electrical equipment. Again, the building partitions are drawn to scale as are such electrical items as fluorescent lighting fixtures, panelboards, and switchgear. The location of other electrical outlets and similar components are only approximated on the drawings because they have to be exaggerated.

3.   Power-riser diagrams to show the service-entrance and panelboard components.

4.   Control wiring schematic diagrams.

5.   Schedules, notes, and large-scale details on construction drawings.

In order to be able to "read" electrical as well as other types of drawings, one must become familiar with the meaning of symbols, lines, and abbreviations used on the drawings and learn how to interpret the message conveyed by the drawings.

Summary

When the construction of a building is contemplated and an architect is hired, a preliminary conference is normally held to decide on specific details for the building. Once this initial information is obtained, the architect and his staff will research the project and commence the design. This design is then conveyed to the workmen by means of working drawings and written specifications.

Questions

1.   What is the name given to the first sketches prepared by an architect?
2.   What is the first page of a set of blueprints normally called?
3.   Natural ground slope is indicated on site plans by _____ lines.
4.   What is a bench mark?
5.   Name some of the utility services shown on plot plans.
6.   What type of plan, more so than any other type of drawing, shows how the building is laid out and is the most readable to the nonprofessional?

7. How do cross sections of a building appear?
8. What term is given to larger-scaled drawings with greater detail?
9. What are footings?
10. What systems do mechanical drawings cover?

Answers to Questions

1. preliminary drawings
2. title sheet
3. grade
4. A permanent object in an area with a predetermined elevation in feet above sea level.
5. water, telephone, electricity, cable TV, and gas and sewer lines
6. floor plans
7. As if they were sliced open to reveal construction details inside of partitions, etc.
8. construction detail drawing
9. Concrete "feet" placed in the ground on which the foundation and subsequent building load is placed.
10. plumbing, heating, ventilation, and air-conditioning

3 Drawing Symbols

Objectives

To introduce symbols used on working drawings for building construction, and other basic essentials necessary for those who must read and interpret construction documents.

The purpose of a working drawing—as applied to the building construction industry—is to show how a certain object, piece of equipment, or system is to be constructed, installed, modified, or repaired.

In the preparation of working drawings for the building construction industry, symbols are used to simplify the work of those preparing the drawings and also to keep the size and bulk of the construction documents to a workable minimum. Can you imagine the time it would take to prepare a set of working drawings for even a simple building if every component had to be drawn as it would be seen with the eye? Consider also the size and number of pages that would be required to hold all of these detailed works of art. Therefore, symbols used on construction drawings may be considered time-saving devices that convey detailed information from the engineer's or architect's designs to the workmen on the job.

Most architects and engineers use symbols adopted by the United States of America Standards Institute (USASI) for use on construction drawings. However, many designers and draftsmen frequently modify these symbols to suit their own particular requirements for the type of projects they normally encounter. For this reason, most drawings have a symbol list or legend drawn and lettered either on each set of working drawings or in the written specifications.

Modified symbols are normally selected by the architect or engineer because they are easier to draw by their draftsmen, are easier for the workmen to interpret, and are sufficient for most applications.

Symbols For Materials

The three most common symbols used to indicate earth, rock, and stone fill are shown in Fig. 3-1; an example of their practical use is

Fig. 3-1. Symbols used to indicate earth, rock and stone fill.

shown in Fig. 3-2. Notice that the earth area in Fig. 3-2 is not entirely marked with the earth symbol. Rather, only the areas around other materials are indicated by the symbol.

Symbols for concrete and related materials such as concrete block, terrazzo, etc. are shown in Fig. 3-3. Examples of their use appear in Fig. 3-4.

Several different symbols are used to indicate metal, and the type employed is usually governed by the size of the drawing or the scale to which the drawing is drawn. When the scale used is too small to indicate the type of metal by symbols, notes are normally provided near the items indicating the exact type of metal to use. A few metal symbols are shown in Fig. 3-5.

Wood symbols appear in Fig. 3-6 and again, the size of the drawing determines, to a great extent, the type of symbol used.

Symbols commonly used to indicate stone and brick are shown in Fig. 3-7, with practical applications of some of the symbols shown in Fig. 3-8.

Glass, gypsum, and miscellaneous architectural symbols can be seen in Fig. 3-9, with practical applications of each in Fig. 3-10.

All of the above symbols are normally used on plan and section drawings as shown in Fig. 3-11, while symbols used in building elevation drawings are shown in Fig. 3-12. Previously described symbols are used when the need arises. Other material combinations are shown in Fig. 3-12 (plan views of exterior walls) and Fig. 3-13 (sections of floor finish).

Most other architectural symbols are pictures of the object depicted as one would see it when looking at it from the angle shown on the drawing. For example, the plan view in Fig. 3-14 shows the walls, partitions, doors, windows, and other details as if the building were horizontally sliced in two with the top section lifted off. Therefore, the drawing actually shows sections of the building as it would look when viewing the sliced open section from above. All doors are shown with their respective swing, window frames, glass, partitions with their finishes—all are shown in great detail on this drawing. Then, using the symbol list studied previously, the persons reading the drawing can tell with what material each section of the building is constructed. When symbols are not appropriate, notes are used to indicate the finish or material.

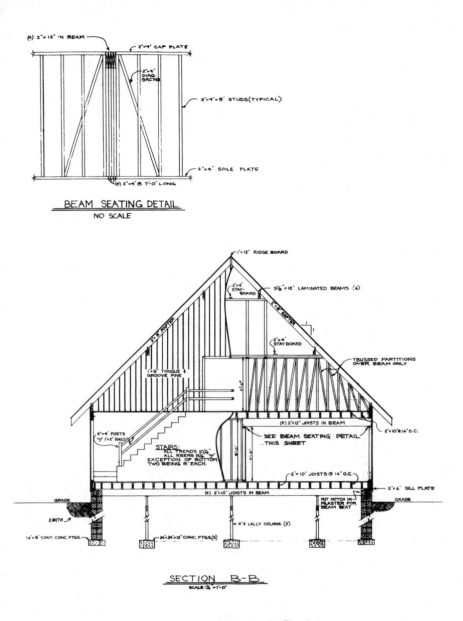

BEAM SEATING DETAIL
NO SCALE

SECTION B-B
SCALE ¼"=1'-0"

Fig. 3-2. Practical use of the symbols shown in Fig. 3-1.

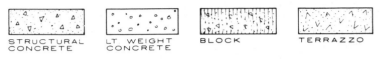

STRUCTURAL CONCRETE   LT WEIGHT CONCRETE   BLOCK   TERRAZZO

Fig. 3-3. Symbols for concrete, concrete block, terrazo, and related materials.

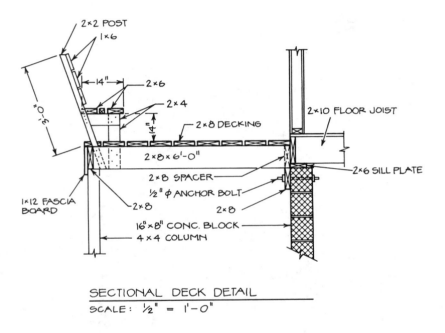

SECTIONAL DECK DETAIL
SCALE: ½" = 1'-0"

Fig. 3-4. Practical application of the symbols shown in Fig. 3-3.

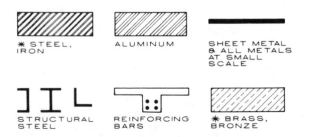

✳ STEEL, IRON   ALUMINUM   SHEET METAL & ALL METALS AT SMALL SCALE

STRUCTURAL STEEL   REINFORCING BARS   ✳ BRASS, BRONZE

Fig. 3-5. Some symbols used to show different types of metal.

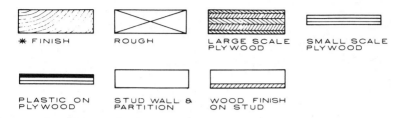

Fig. 3-6. Typical wood symbols used on construction drawings.

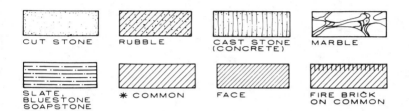

Fig. 3-7. Symbols commonly used to indicate stone and brick.

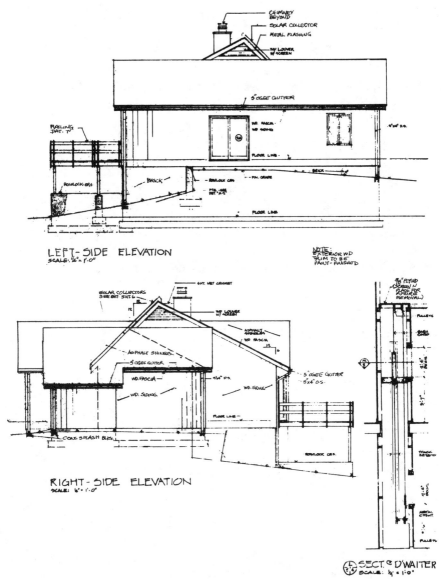

Fig. 3-8.   Practical application of the symbols shown in Fig. 3-7.

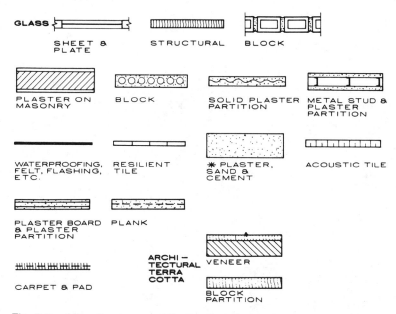

GLASS — SHEET & PLATE   STRUCTURAL   BLOCK

PLASTER ON MASONRY   BLOCK   SOLID PLASTER PARTITION   METAL STUD & PLASTER PARTITION

WATERPROOFING, FELT, FLASHING, ETC.   RESILIENT TILE   *PLASTER, SAND & CEMENT   ACOUSTIC TILE

PLASTER BOARD & PLASTER PARTITION   PLANK

CARPET & PAD   ARCHI — TECTURAL TERRA COTTA   VENEER

BLOCK PARTITION

Fig. 3-9.   Miscellaneous architectural symbols.

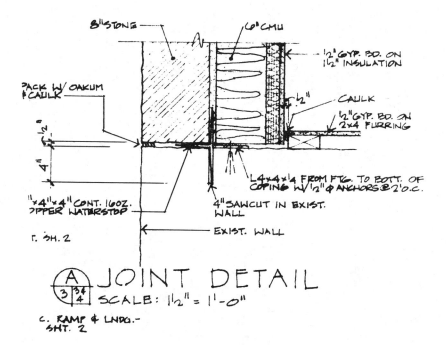

8" STONE   6" CMU

1/2" GYP. BD. ON 1 1/2" INSULATION

PACK W/ OAKUM & CAULK

CAULK

1/2"

1/2" GYP. BD. ON 2×4 FURRING

L 4×4×1/4 FROM FTG. TO BOTT. OF COPING W/ 1/2" ⌀ ANCHORS @ 2'0.c.

1/2"×4"×4" CONT. 16OZ. COPPER WATERSTOP

4" SAWCUT IN EXIST. WALL

r. SH. 2

EXIST. WALL

Ⓐ JOINT DETAIL
3 | 3/4
4   SCALE: 1 1/2" = 1'-0"

c. RAMP & LNDG.- SHT. 2

Fig. 3-10.   Practical application of glass and gypsum board symbols.

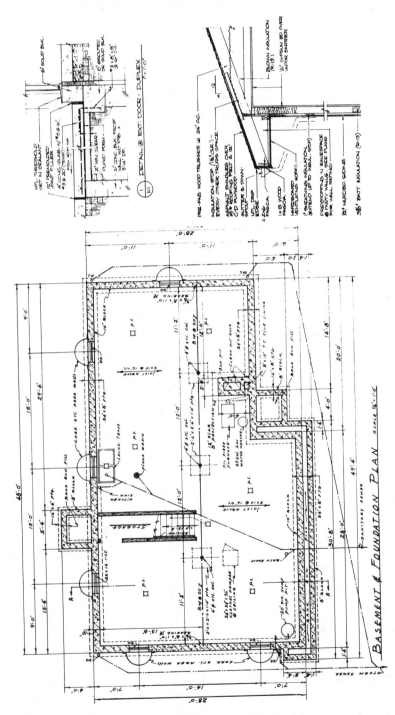

Fig. 3-11. Combination of symbols used on plan and section drawings.

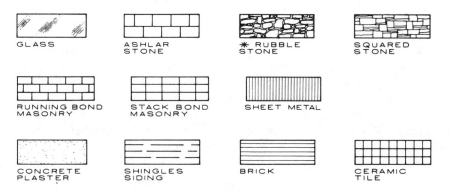

GLASS

ASHLAR STONE

✳ RUBBLE STONE

SQUARED STONE

RUNNING BOND MASONRY

STACK BOND MASONRY

SHEET METAL

CONCRETE PLASTER

SHINGLES SIDING

BRICK

CERAMIC TILE

Fig. 3-12. Combination of symbols used on elevation drawings.

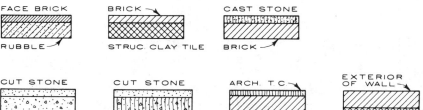

FACE BRICK

RUBBLE

BRICK

STRUC. CLAY TILE

CAST STONE

BRICK

CUT STONE

STRUC. CONC.

CUT STONE

CONCRETE BLOCK

ARCH. T.C.

BRICK

EXTERIOR OF WALL

INTERIOR

Fig. 3-13. Symbol combination used on plan view of exterior walls.

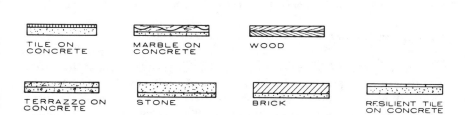

TILE ON CONCRETE

MARBLE ON CONCRETE

WOOD

TERRAZZO ON CONCRETE

STONE

BRICK

RESILIENT TILE ON CONCRETE

Fig. 3-14. Symbols used to indicate floor finish.

Electrical Symbols

Electrical drawings—like any other type of building construction drawing—must be prepared by electrical draftsmen in a given time period in order to stay within an allotted budget. Therefore, symbols are also used on electrical drawings to simplify the work, for both the draftsmen and the workers interpreting the drawings.

Fig. 3-15 shows a list of electrical symbols currently in use on most electrical drawings. This list represents a good set of electrical symbols in that they are:

- easy to draw
- easily interpreted by workmen
- sufficient for most applications

It is evident from the list of symbols in Fig. 3-15 that many symbols have the same basic form, but their meanings differ slightly because of the addition of a line, mark, or abbreviation. Therefore, a good procedure to follow in learning the different electrical symbols is to first learn the basic form and then apply the variations of that form to obtain the different meanings.

Note that some of the symbols listed contain abbreviations, such as WT for *watertight* and S for *switch*. Others are simplified pictographs, such as the symbol for safety switch or the symbol for flush-mounted panelboard.

In learning these symbols (and other symbols for that matter), it is a good idea to obtain a catalog from one of the electrical supply companies and compare the actual appearance of the component or piece of equipment with the symbol.

Plumbing Symbols

In preparing a plumbing drawing, all pipe, fittings, fixtures, valves and other components are shown by symbols such as those listed in Fig. 3-16. The use of these symbols simplifies considerably the preparation of piping drawings and conserves a great deal of time and effort.

Heating, Ventilating, and Air-Conditioning Symbols

The main purpose of HVAC drawings is to show the location of the heating, cooling, and air-conditioning units along with their related duct work and piping. Graphical symbols on HVAC drawings are similar in pattern to those used for plumbing. The list of symbols shown in Fig. 3-17 are typical of those in current use.

NOTE: THESE ARE STANDARD SYMBOLS AND MAY NOT ALL APPEAR ON THE PROJECT DRAWINGS; HOWEVER, WHEREVER THE SYMBOL ON PROJECT DRAWINGS OCCURS, THE ITEM SHALL BE PROVIDED AND INSTALLED.

	FLUORESCENT STRIP		CONDUIT, CONCEALED IN CEILING OR WALL
	FLUORESCENT FIXTURE		CONDUIT, CONCEALED IN FLOOR OR WALL
	INCANDESCENT FIXTURE, RECESSED		CONDUIT, EXPOSED
	INCANDESCENT FIXTURE, SURFACE OR PENDANT		FLEXIBLE METALLIC ARMORED CABLE
	INCANDESCENT FIXTURE, WALL-MOUNTED		HOME RUN TO PANEL - NUMBER OF ARROWHEADS INDICATES NUMBER OF CIRCUITS. NOTE:
	LETTER "E" INSIDE FIXTURES INDICATES CONNECTION TO EMERGENCY LIGHTING CIRCUIT		ANY CIRCUIT WITHOUT FURTHER DESIGNATION INDICATES A TWO-WIRE CIRCUIT. FOR A GREATER NUMBER OF WIRES, READ AS
	NOTE: ON FIXTURE SYMBOL, LETTER OUTSIDE DENOTES SWITCH CONTROL		FOLLOWS - 3 WIRES, 4 WIRES, ETC.
	EXIT LIGHT, SURFACE OR PENDANT	—T—	TELEPHONE CONDUIT
		—TV—	TELEVISION-ANTENNA CONDUIT
	EXIT LIGHT, WALL-MOUNTED	—S—	SOUND-SYSTEM CONDUIT - NUMBER OF CROSSMARKS
	INDICATES FIXTURE TYPE		INDICATES NUMBER OF PAIRS OF CONDUCTORS.
	RECEPTACLE, DUPLEX-GROUNDED	F	FAN COIL-UNIT CONNECTION
	RECEPTACLE, WEATHERPROOF		MOTOR CONNECTION
	COMBINATION SWITCH AND RECEPTACLE	M.H.	MOUNTING HEIGHT
	RECEPTACLE, FLOOR-TYPE	F	FIRE-ALARM STRIKING STATION
		G	FIRE-ALARM GONG
	RECEPTACLE, POLARIZED (POLES AND AMPS INDICATED)	D	FIRE DETECTOR
s	SWITCH, SINGLE-POLE	SD	SMOKE DETECTOR
s₃,₄	SWITCH, THREE-WAY, FOUR-WAY	B	PROGRAM BELL
sₚ	SWITCH AND PILOT LIGHT	Y	YARD GONG
sₜ	SWITCH, TOGGLE W/ THERMAL OVERLOAD PROTECTION	C	CLOCK
	PUSH BUTTON	M	MICROPHONE, WALL-MOUNTED
	BUZZER	M	MICROPHONE, FLOOR-MOUNTED
	LIGHT OR POWER PANEL		SPEAKER, WALL-MOUNTED
	TELEPHONE CABINET	S	SPEAKER, RECESSED
J	JUNCTION BOX	V	VOLUME CONTROL
	DISCONNECT SWITCH - FSS - FUSED SAFETY SWITCH. NFSS - NONFUSED SAFETY SWITCH		TELEPHONE OUTLET, WALL
			TELEPHONE OUTLET, FLOOR
	STARTER		TELEVISION OUTLET
A.F.F.	ABOVE FINISHED FLOOR		

Fig. 3-15.   Electrical symbols currently in use.

MECHANICAL DRAWING SYMBOLS

NOTE: THESE ARE STANDARD SYMBOLS AND MAY NOT ALL APPEAR ON THE PROJECT DRAWINGS; HOWEVER, WHEREVER THE SYMBOL ON THE PROJECT DRAWINGS OCCURS, THE ITEM SHALL BE PROVIDED AND INSTALLED.

Symbol	Description	Symbol	Description
—S—	STEAM PIPE	MBH	THOUSAND BTU PER HOUR
—C—	CONDENSATE RETURN PIPE	GPM	GALLONS PER MINUTE
—HWS—	HOT WATER SUPPLY PIPE	CFM	CUBIC FEET PER MINUTE
—HWR—	HOT WATER RETURN PIPE	φ	ROUND
—CWS—	CHILLED WATER SUPPLY PIPE	φ	SQUARE
—CWR—	CHILLED WATER RETURN PIPE	SA	SUPPLY AIR
—HCS—	COMB HOT-CHILLED WATER SUPPLY	RA	RETURN AIR
—HCR—	COMB HOT-CHILLED WATER RETURN	OA	OUTSIDE AIR
—CS—	CONDENSER WATER SUPPLY PIPE	EA	EXHAUST AIR
—CR—	CONDENSER WATER RETURN PIPE	HSWR	HIGH SIDEWALL REGISTER
—D—	DRAIN PIPE FROM COOLING COIL	HSWG	HIGH SIDEWALL GRILLE
—FOS—	FUEL OIL SUPPLY PIPE	LSWR	LOW SIDEWALL REGISTER
—FOR—	FUEL OIL RETURN PIPE	LSWG	LOW SIDEWALL GRILLE
—R—	REFRIGERANT PIPE	CSR	CEILING SUPPLY REGISTER
—o	PIPE RISING	CR	CEILING REGISTER
—o	PIPE TURNING DOWN	CG	CEILING GRILLE
UNION	UNION	FR	FLOOR REGISTER
REDUCER - CONCENTRIC	REDUCER - CONCENTRIC	FG	FLOOR GRILLE
REDUCER - ECCENTRIC	REDUCER - ECCENTRIC	CD	CEILING DIFFUSER
STRAINER	STRAINER	TV	TURNING VANES
GATE VALVE	GATE VALVE	AE	AIR EXTRACTOR
GLOBE VALVE	GLOBE VALVE	SD	SPLITTER DAMPER
VALVE IN RISER	VALVE IN RISER	MD	MANUAL DAMPER
CHECK VALVE	CHECK VALVE	FD	FIRE DAMPER
PRESSURE REDUCING VALVE	PRESSURE REDUCING VALVE	DL	DUCT LINER IN DUCT
PRESSURE RELIEF VALVE	PRESSURE RELIEF VALVE	AHU	AIR HANDLING UNIT
SQUARE HEAD COCK	SQUARE HEAD COCK	BU	BLOWER UNIT
BALANCING VALVE	BALANCING VALVE	FCU	FAN COIL UNIT
3-WAY CONTROL VALVE	3-WAY CONTROL VALVE	HWC	HOT WATER CONVECTOR
2-WAY CONTROL VALVE	2-WAY CONTROL VALVE	UV	UNIT VENTILATOR
PITCH PIPE MINIMUM 1"/40'	PITCH PIPE MINIMUM 1"/40'	WH	WALL HEATER
ANCHOR LOCATION	ANCHOR LOCATION	UH	UNIT HEATER
FLEXIBLE PIPE CONNECTION	FLEXIBLE PIPE CONNECTION	WF	WALL FIN RADIATION
IN-LINE PUMP	IN-LINE PUMP	PRV	POWER ROOF VENTILATOR
BOTTOM TAKE-OFF	BOTTOM TAKE-OFF	UVS	UTILITY VENT SET
TOP TAKE-OFF	TOP TAKE-OFF	PF	PROPELLER FAN
		T	THERMOSTAT
PRESSURE GAGE	PRESSURE GAGE	T_N	NIGHT THERMOSTAT
THERMOMETER	THERMOMETER	T_H	THERMOSTAT - HEATING ONLY
HOT WATER RISER	HOT WATER RISER	T_C	THERMOSTAT - COOLING ONLY
CHILLED WATER RISER	CHILLED WATER RISER	T—o	THERMOSTAT - REMOTE BULB
FAN COIL UNIT	FAN COIL UNIT	+6'-7"	MOUNTING HEIGHT ABOVE FINISHED FLOOR
EQUIPMENT AS INDICATED	EQUIPMENT AS INDICATED	NIC	NOT IN CONTRACT
AIR INTO REGISTER	AIR INTO REGISTER		SUPPLY AIR DUCT SECTION
AIR OUT OF REGISTER	AIR OUT OF REGISTER		
AIR FLOW THRU UNDERCUT OR	AIR FLOW THRU UNDERCUT OR		RETURN OR EXHAUST DUCT SECTION
LOUVERED DOOR	LOUVERED DOOR		
TURNING VANES	TURNING VANES		FLEXIBLE DUCT CONNECTION
AIR EXTRACTOR	AIR EXTRACTOR		

Fig. 3-16. Plumbing symbols used on most construction drawings.

MECHANICAL   SYMBOLS

HWS	HOT WATER SUPPLY
HWR	HOT WATER RETURN
HSWG	HIGH SIDEWALL GRILLE
CD	CEILING DIFFUSER
CR	CEILING REGISTER
CG	CEILING GRILLE
MD	MANUAL DAMPEF
FD	FIRE DAMPER
UH	UNIT HEATER
PRV	POWER ROOF VENTILATOR
⊕	THERMOSTAT
MBH	THOUSAND BTU PER HOUR
CFM	CUBIC FEET PER MINUTE
OA	OUTSIDE AIR
⟶	AIR OUT OF REGISTER
 ⟶	AIR INTO REGISTER
 ⟶Ⓐ⟶	LOUVERED OR UNDERCUT DOOR
 ⟶	PITCH DOWN AS DIRECTED

Fig. 3-17. Graphic symbols used on heating, ventilating, and air conditioning drawings.

Fig. 3-18. Drawing showing duct work layout.

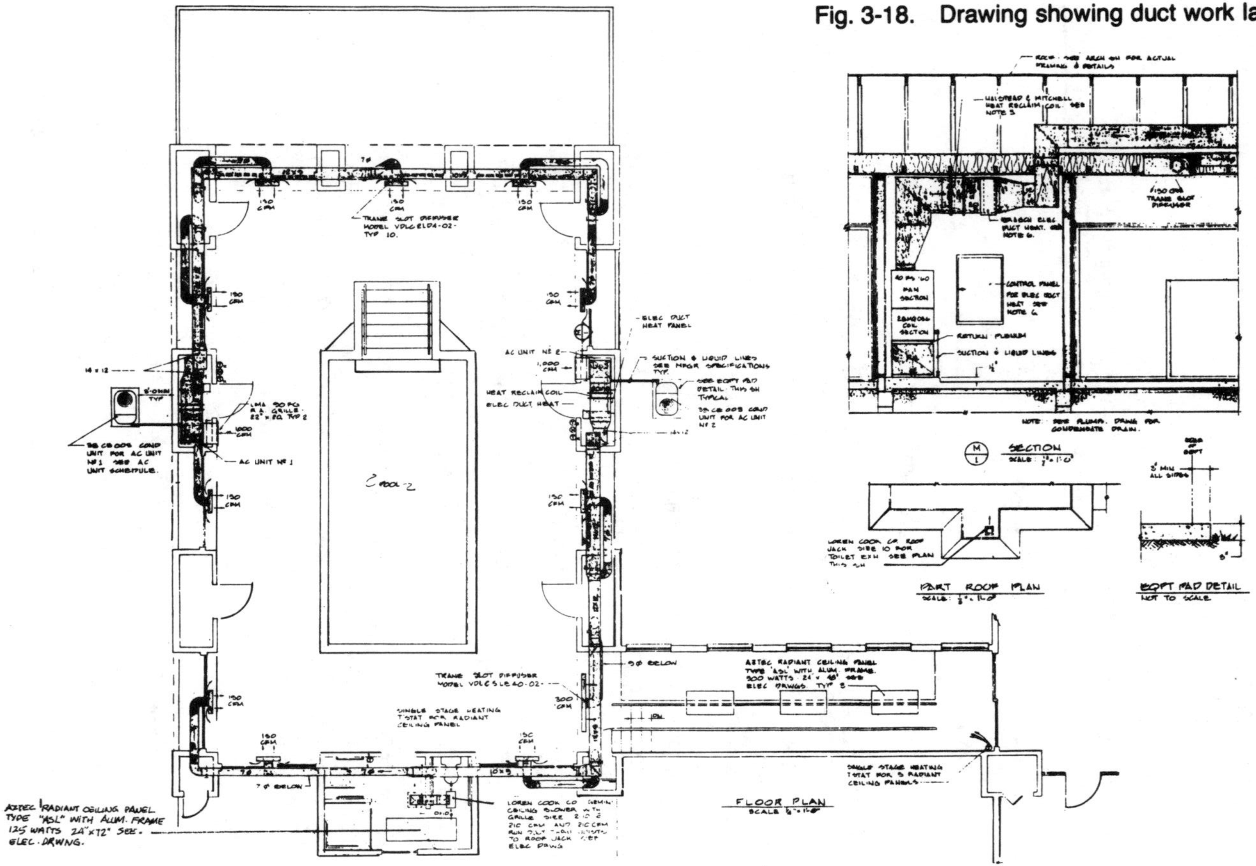

Notice that the ductwork shown on the drawing in Fig. 3-18 uses two types of symbols. The main duct consists of a rectangle to represent the actual size of the ductwork—including reductions. However, the branches are shown by a single line with the dimensions of each noted adjacent to these lines. You will find both methods used on HVAC drawings throughout the building construction industry—either singly or in combination. Obviously, the two-line method is easier to use since the ductwork appears as it would if the viewer were looking down on the system from above.

Summary

Drawing symbols are used on working drawings to simplify the work of draftsmen and to keep the size and bulk of the drawings to a workable minimum. In short, they may be considered as time-saving devices that convey the design to the workmen and others involved in the building construction project.

Most architects and engineers use symbols adopted by the United States of America Standards Institute for use on construction drawings. Some drawings, however, deviate from the standard symbols, but usually contain legends or symbol lists to identify any symbol used.

Questions

1. Draw the three most common symbols used to indicate earth.
2. Draw the symbol commonly used to indicate a standard electrical duplex receptacle.
3. Draw one symbol used to identify any type of metal.
4. Shown in Figure 3–19 are 18 symbols commonly found on building construction drawings. In the space provided, place the number corresponding to the correct answer found in the list.

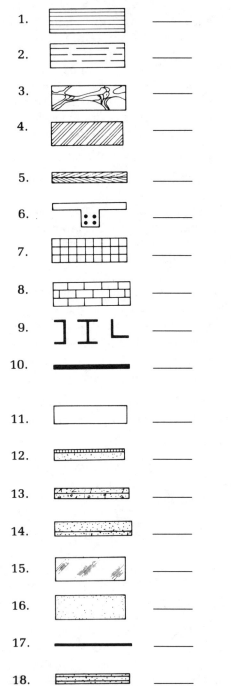

1. _____   a. running bond masonry

2. _____   b. tile on concrete

3. _____   c. marble

4. _____   d. waterproofing, felt, flashing, etc.

5. _____   e. structured steel

6. _____   f. reinforcing bars

7. _____   g. brick

8. _____   h. concrete plaster

9. _____   i. aluminum

10. _____   j. plaster board and plaster partition

11. _____   k. shingles siding

12. _____   l. ceramic tile

13. _____   m. wood

14. _____   n. stud wall and partition

15. _____   o. terrazzo on concrete

16. _____   p. stone

17. _____   q. glass

18. _____   r. sheet metal

Fig. 3-19.   Symbols for question #4 of Questions.

4　Orthographic Projections

Objectives

To explain an orthographic-projection drawing, to learn the principal parts of an architectural drawing, and to learn the use of the Architect's Scale.

Orthographic Projection

Many types of drawings are used to prepare working drawings for building construction. However, the views known as an orthographic projection will be the most commonly seen on construction projects. The reason for this is that although pictorial views of objects give a realistic appearance, only drawings with exact information concerning shape and size and material can be used to properly construct an object or a system. This type of information is usually best given in a drawing where several related views of an object are presented in the proper manner.

The pictorial drawing of a cube in Fig. 4-1 is a single-view drawing that shows how the object appears to the viewer. The words length, width, and height are used to show the basic dimensions on the drawings. If this same object were converted to an orthographic projection, it would appear as shown in Fig. 4-2. Here, only three separate views are required, since the top and bottom are the same dimensions, the back and front are the same, and both sides are the same.

In both figures (Figs. 4-1 and 4-2) the width represents the horizontal distances on the drawing in the top and front views. The height refers to the vertical distances on the drawing in the front and side views, while the depth refers to the vertical distances in the top view and the horizontal distances in the side view.

An orthographic projection of a building would consist of the views shown in Fig. 4-3. One drawing shows the building as though the observer were looking straight at the front; one as though the observer were looking straight at the left side, one straight at the right side, and one as though the observer were looking straight at the rear of the building. These drawings may be referred to as *building elevations*.

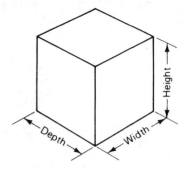

Fig. 4-1.   Pictorial drawing of a cube.

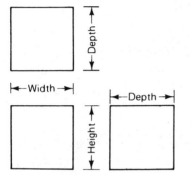

Fig. 4-2.   Orthographic projection of the cube in Fig. 4-1.

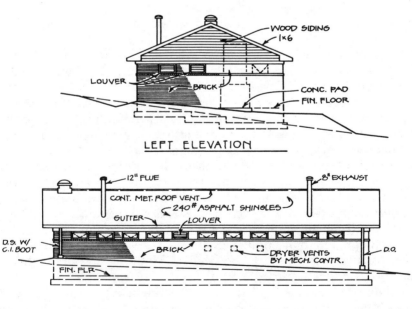

Fig. 4-3.   Orthographic projection of a building (building elevation).

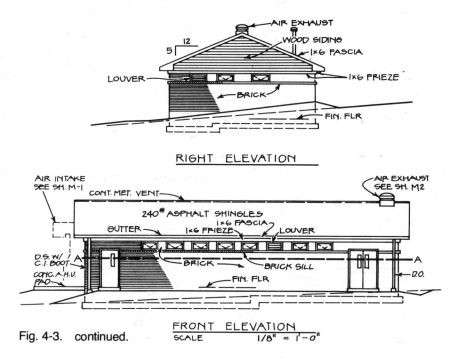

Fig. 4-3. continued.

Referring again to Fig. 4-3, assume that a horizontal cut is made through the building along a line indicated as "A-A" on the front elevation. Then imagine that the top part is removed and a drawing is then made as if the artist were looking straight down at the remaining part. This drawing is known as a floor plan. It indicates the outside wall lines, interior partitions, windows, doors, and similar details. The final drawing would appear as shown in Fig. 4-4.

If the building should have more than one floor, horizontal cuts may be taken at varying distances from the ground in order to show the other floors on the drawing. For example, the elevation drawings of the building in Fig. 4-5 show three different floor levels. With the horizontal cuts made at "A-A", "B-B", and "C-C" the plans of the three separate floors would appear in Figs. 4-6, 4-7 and 4-8 respectively.

Using the Architect's Scale

Drawings of buildings and related equipment obviously cannot be drawn to full scale on working drawings. Therefore, the drawing is reduced in size so that all the distances on the drawing are drawn smaller than the actual dimensions of the building—all dimensions

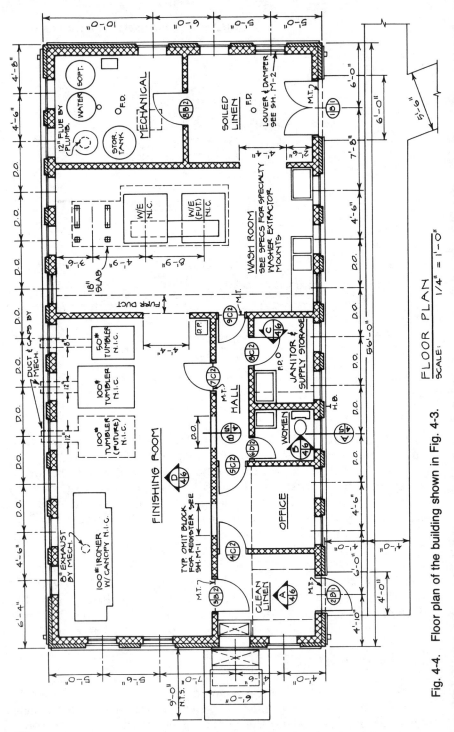

Fig. 4-4.   Floor plan of the building shown in Fig. 4-3.

58

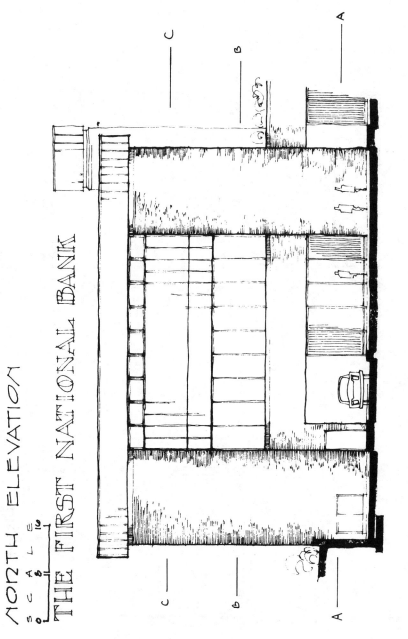

Fig. 4-5. Elevation drawings of a 3-story building.

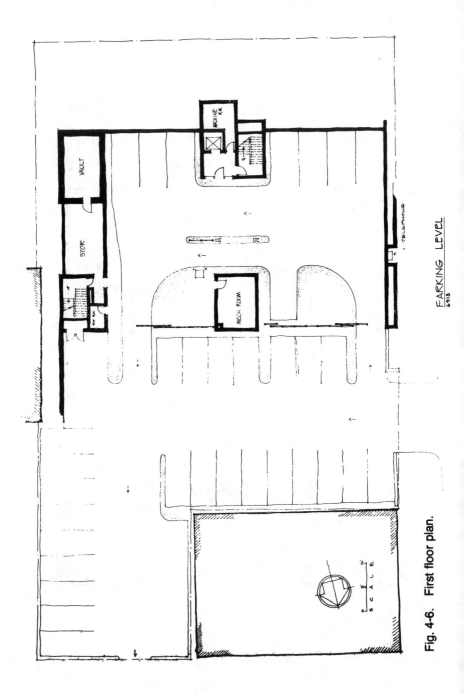

Fig. 4-6. First floor plan.

60

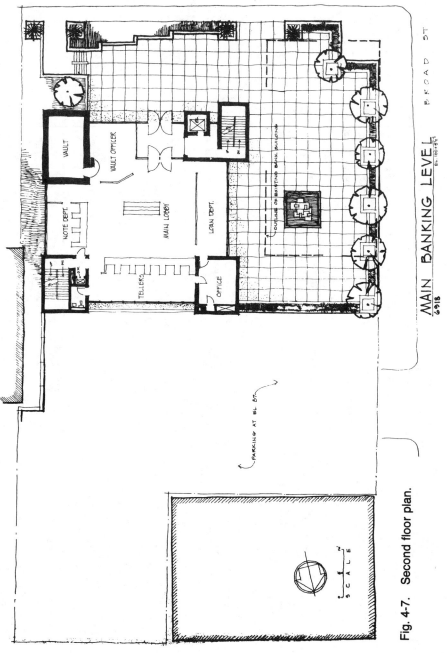

Fig. 4-7. Second floor plan.

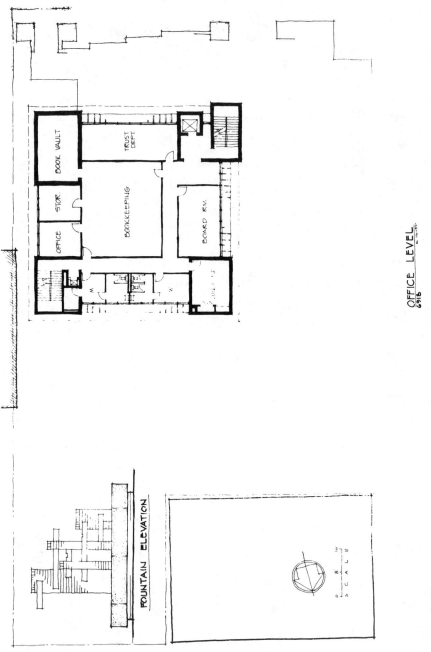

OFFICE LEVEL

FOUNTAIN ELEVATION

SCALE

Fig. 4-8.   Third floor plan.

62

being reduced in the same proportion. The ratio, or relation, between the size of the drawing and the size of the object drawn is indicated on the drawing paper ($\frac{1}{4}$" = 1' - 0" for example), and the dimensions are the actual dimensions of the building or other object, not the distance measured on the drawing.

The most common method of reducing all the dimensions in the same proportion is to select a certain distance and let that distance represent 1 foot. This distance is then divided into 12 equal parts, each of which represents one inch. If $\frac{1}{2}$-inch divisions are required for the drawing, these twelfths are further subdivided into halves. This method gives a usable scale that represents the common foot rule known as the architect's scale; when a measurement is taken on the building or object itself, it is made with the standard foot rule; when a measurement is laid off on the drawing, it is made with the reduced foot rule known as the architect's scale.

Architect's scales are available in many different degrees or scales; that is, 1 inch = 1 foot, $\frac{1}{4}$ inch = 1 foot, etc.

Figure 4-9 shows part of a building floor plan drawn to a scale of $\frac{1}{4}$ inch = 1 foot. The dimensions in question are found by placing the $\frac{1}{4}$-inch architect's scale on the drawing (as shown) and reading the figures.

An architect's scale showing four degrees ($\frac{1}{8}$" = 1', $\frac{1}{4}$" = 1', $\frac{1}{2}$" = 1', and 1" = 1') is shown in Fig. 4-10. The dimensions of the various lines are as follows: A equals 12 feet 6 inches on the $\frac{1}{8}$-inch scale; B equals 8 feet 6 inches and C equals 2 feet 6 inches—both on the $\frac{1}{4}$-inch scale; D equals 1 foot 9 inches on the $\frac{1}{2}$-inch scale; E equals 5 feet 0 inches on the 1-inch scale, and F equals 4 inches on the 1-inch scale.

Every drawing prepared in a given reduced scale should be plainly marked as to what the scale is. For example, the building floor plan in Fig. 4-11 should be marked

FIRST FLOOR PLAN
Scale: $\frac{1}{4}$" = 1'0"

Architectural Dimensioning

The common rules that apply to other types of drawings also apply to architectural drawings. The dimensions given must be clear and definite, and must tell the workmen the exact sizes of all parts of the building. Furthermore, they must check with one another from place to place and from plan to elevation to section, etc. Several points should be observed in an architectural dimensioning:

1. Keep all outside dimension lines well away from the building lines. They should be located a minimum of $\frac{3}{4}$ in. from the building lines and should be approximately $\frac{5}{16}$ to $\frac{3}{8}$ inch apart.

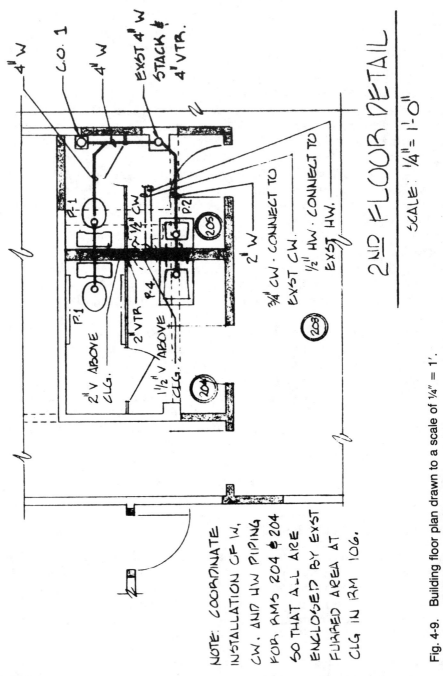

NOTE: COORDINATE INSTALLATION OF W, CW. AND HW PIPING FOR RMS 204 & 204 SO THAT ALL ARE ENCLOSED BY EXST FURRED AREA AT CLG IN RM 106.

4" W
C.O. 1
4" W
EXST 4" W STACK & 4" VTR.

P-1
P-3
2" 1/2 CW
P-2
206

2" V ABOVE CLG.
P-1
P-4
2" VTR
1 1/2" V ABOVE CLG.
204

2" W
3/4" CW - CONNECT TO EXST CW.
1/2" HW - CONNECT TO EXST HW.
208

2ND FLOOR DETAIL

SCALE: 1/4" = 1'-0"

Fig. 4-9.   Building floor plan drawn to a scale of 1/4" = 1'.

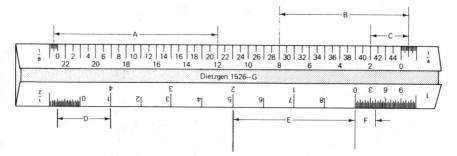

Fig. 4-10.   Architect's scale showing 4 degrees.

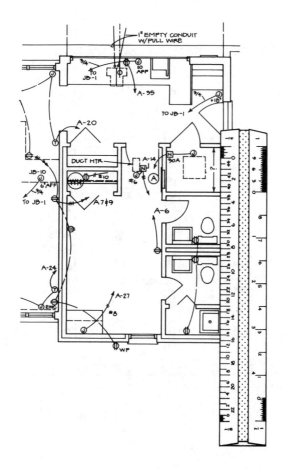

Fig. 4-11.   Building floor plan drawn to a scale of ¼" = 1'.

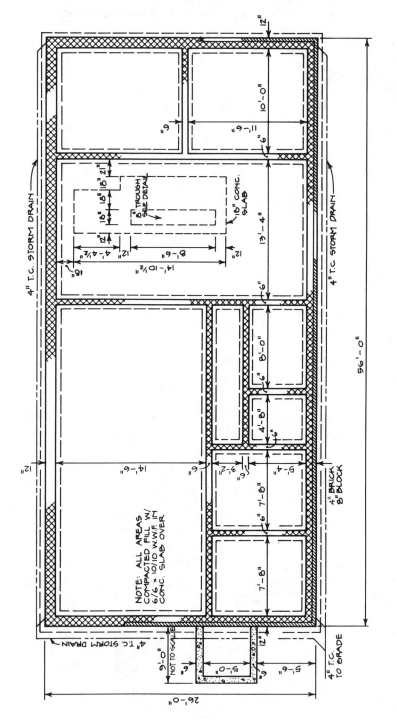

Fig. 4-12.   One method of dimensioning architectural drawings.

2. Masonry openings should be marked "M.O." when an exact opening is required. When the opening is for a window or a door, a nominal dimension may be used.

3. Dimensions are usually to the centerlines of partitions or are made to the outside walls. In any case, the wall thicknesses should also be shown.

4. Dimensions should be provided to the centerlines of columns in both directions.

5. Dimensions for openings are normally made to the centerline of the opening or to the sides of the opening, as required.

Figures 4-12 and 4-13 show various methods of dimensioning architectural drawings. Notice that most of the dimensions on these drawings are from center to center of various objects. The reason is that lumber and other construction materials often vary slightly in size, making it impossible to indicate actual values of edges of structural members. This way, if structural members vary somewhat in size, the true location is always achieved.

Besides dimension lines and numerals, every drawing should include a plainly marked scale to which the drawing is made. The scale is usually inserted as part of, or adjacent to, the title block or else shown directly beneath a drawing title elsewhere on the sheet.

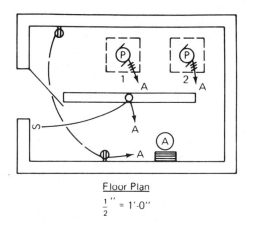

Floor Plan

$\frac{1}{2}'' = 1'\text{-}0''$

Fig. 4-13.   Another method of dimensioning architectural drawings.

The Metric System

The metric system is an international language of measurement. Its symbols are identical in all languages. Just as the English language is governed by rules of spelling, punctuation, and pronunciation, so is the language of measurement. Uniformity of usage facilitates comprehension, and minimizes the chance of decimal point errors.

This section covers the editorial rules of the metric system, and is designed to serve as a convenient reference guide. It uses the modern form of the metric system, called "The International System of Units" (abbreviated SI). A limited number of the units outside SI that are acceptable for use with SI are also included.

The metric system is based on the meter, which is equal to 39.370432 inches. The value commonly used is 39.37 inches and is authorized by the U.S. Government.

There are three principal units—the meter, the liter (pronounced "lee-ter"), and the gram, the units of length, capacity, and weight, respectively. Multiples of these units are obtained by prefixing to the names of the principal units the Greek words deka (10), hecto (100), and kilo (1,000); the submultiples, or divisions, are obtained by prefixing the Latin words deci ($\frac{1}{10}$), centi ($\frac{1}{100}$), and milli ($\frac{1}{1000}$). These prefixes form the key to the entire system.

MEASURES OF LENGTH

10 millimeters (mm)	= 1 centimeter	cm
10 centimeters	= 1 decimeter	dm
10 decimeters	= 1 meter	m
10 meters	= 1 dekameter	dam
10 decameters	= 1 hectometer	hm
10 hectometers	= 1 kilometer	km

MEASURES OF SURFACE
(Not Land)

100 square millimeters (mm^2)	= 1 square centimeter	cm^2
100 square centimeters (cm^2)	= 1 square decimeter	dm^2
100 square decimeters (dm^2)	= 1 square meter	m^2

MEASURES OF VOLUME

1000 cubic millimeters (mm³)	= 1 cubic centimeter	........cm³
1000 cubic centimeters	= 1 cubic decimeter	........dm³
1000 cubic decimeters	= 1 cubic meter	.............m³

MEASURES OF CAPACITY

10 milliliters (ml)	= 1 centiliter	cl
10 centiliters	= 1 deciliter	................dl
10 deciliters	= 1 liter	l
10 liters	= 1 dekaliter	...............dal
10 dekaliters....................	= 1 hectoliter	...............hl
10 hectoliters	= 1 kiloliter	.................kl

The liter is equal to the volume occupied by 1 dm³.

MEASURES OF WEIGHT

10 milligrams (mg)	= 1 centigram	................cg
10 centigrams	= 1 decigram	...............dg
10 decigrams	= 1 gram	.......................g
10 grams	= 1 dekagram	...............dg
10 dekagrams....................	= 1 hectogram	...............hg
10 hectograms	= 1 kilogram	.................kg
1000 kilograms	= 1 ton	....................ton

The gram is the weight of 1 cm³ of pure distilled water at a temperature of 39.2°F (4°C) the kilogram is the weight of 1 liter of water; the ton is the weight of 1 cm³ of water.

METRIC CONVERSION FACTORS

In order to use the following factors for converting from English to metric units, it is necessary to transform the equations; for example,

1000 km × .621 × 621 mi, but 1000 mi ÷ .621 = 1610 km.
km × .621 = mi
km ÷ 1.609 = mi
km × 3,281 = ft
m × 39.37 = in
m × 3.281 = ft
m × 1.094 = yd
cm × .3937 = in
cm ÷ 2.54 = in
mm × .03937 = in
mm ÷ 25.4 = in

$km^2 \times 247.1 = acre$
$m^2 \times 10.764 = ft^2$
$cm^2 \times .155 = in^2$
$cm^2 \div 6.451 = in^2$
$mm^2 \times .00155 = in^2$
$mm^2 \div 645.1 = in^2$
$m^3 \times 35.315 = ft^3$
$m^3 \times 1.308 = yd^3$
$m^3 \times 264.2 = gal$ (U.S.)
$cm^3 \div 16.383 = in^3$
$l \times 61.022 = in^3$
$l \times .2642 = gal$ (U.S.)
$l \times 3.78 = gal$ (U.S.)
$l \times 28.316 = ft^3$
$g \times 15.432 = gr$
$g \times 981 = dynes$
$g \times 28.35 = oz$ (avoir.)
kilograms per $cm^2 \times 14.22 = lb/in^2$
$Kg \times 2.205 = lb$
$Kg \times 35.3 = oz$ (avoir.)
$Kg \times .0011023 = tons$ (2000 lb)
$Kg/cm^2 \times 14,223 = lb/in^2$
$Kg-m \times 7.233 = ft-lb$
kilowatts (kW) $\times 1.34 = hp$
watts $\times 746 = hp$
watts $\times .7373 = ft-lb$ per sec
joules $\times .7373 = ft-lb$
Calorie (kilogram-degree) $\times 3.968 = Btu$
Calorie (kilogram-degree) $\div .252 = Btu$
joules $\times .24 = gram-calories$
gram-calories $\times 4.19 = joules$
gravity (Paris) $= 981$ cm/s
(Degrees Celsius $\times 1.8$) $+ 32 = $ degrees Fahrenheit
(Degrees Fahrenheit $- 32$) $\div 1.8 = $ degrees Celsius

Summary

An orthographic projection of an object represents the following views: one shows the object as though the observer were looking straight at the front; one as though the observer were looking straight at the left side; one at the right side; and one as though the observer were looking straight at the rear of the building. When an orthographic projection is made of a building, these four drawings may be referred to as building elevations.

Another view is necessary to complete an orthographic projection: a view as if the observer were above the object and looking striaght down on it. In this latter type of drawing, if a horizontal cut were made through the building, and the top portion lifted off, the remaining view would be known as a floor plan.

The most common method of measuring scaled drawings is by using the architect's scale.

The metric system is an internation language of measurement that is based on the meter, which is equal to 39.37 inches.

Questions

1. How many drawings are normally utilized to show the elevation of a building?
2. Name four sections of a building normally shown on a building floor plan.
3. On a drawing drawn to a scale of ¼″ = 1′, how many inches on the drawing would represent 34′?
4. What is the minimum distance that dimension lines should be located from the building lines on a drawing?
5. To what points of partitions are dimensions usually given?
6. Why are dimensions normally given from center to center of various objects?
7. Where is the scale of a drawing normally shown?
8. A meter is equal to how many inches?
9. How many millimeters are in 1 centimeter?
10. How many grams are in 1 decagram?

Answers to Questions

1. 4
2. partitions, wall lines, windows, and doors
3. 8½″
4. ¾″
5. center to center
6. Because building materials vary in size.
7. In the little block or under the drawing title.
8. 39.37
9. 10
10. 10

5 Sectional Views and Construction Details

Objectives

To give the reader instructions on building sections and large-scale details so that he will obtain a better understanding of buildings as they are shown on architectural drawings.

Sectional Views

A section of any object—as applied to construction drawings—is what could be seen if the object were sliced or sawed into two parts at the point where the section was taken. For example, if you wanted to see how a golf ball is constructed, you could place the ball in a vice and saw it in half with a hacksaw. When the two parts are separated, you can easily see how the ball is constructed, or you would at least have a view of its internal construction. This typifies the need for sectional views in building construction drawings.

In dealing with sections, workmen must use a considerable amount of visualization, since there are no given rules for determining what a section will look like. Take a piece of common galvanized water pipe, for example. If the pipe were cut vertically, the section will appear as shown in Fig. 5-1; cut horizontally, the section will appear as that shown in Fig. 5-2; cut on a slant and the section will form an ellipse as shown in Fig. 5-3.

Without sections, all invisible details could not be shown—except for dotted lines, (see Fig. 5-4), which are seldom clear and which make the drawing difficult to read.

The theory of the construction of a sectional view is illustrated in Figs. 5-5 through 5-7. Figure 5-5 shows a pipe viewed from the end (a) and side (b). In Fig. 5-6 (a), a cutting plane is shown passing through the pipe; this is where the section is taken and is indicated by the cutting-plane line. In Fig. 5-6 (b), the portion of the pipe section between the viewer and the cutting plane has been removed to reveal the interior details of the pipe. Then in Fig. 5-7, the cutting plane is removed, and the section would be drawn in an orthographic view.

Fig. 5-1.   Vertical section of a water pipe.

Fig. 5-2.   Horizontal cut of a water pipe.

Fig. 5-3.   Section of a water pipe cut on a slant.

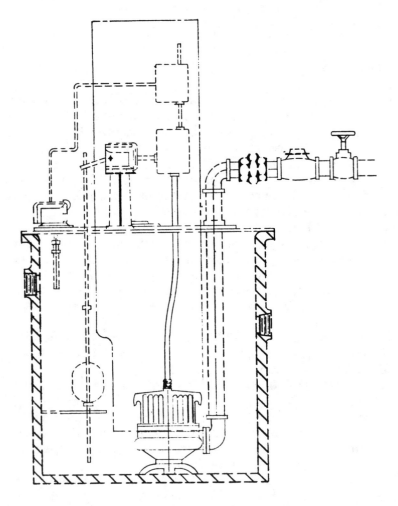

Fig. 5-4.   Invisible detail shown on a drawing utilizing dotted lines.

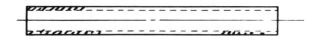

Fig. 5-5. A pipe viewed from the end and side.

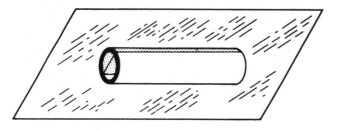

Fig. 5-6A. Cutting plane shown passing through pipe.

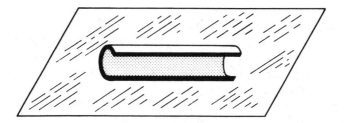

Fig. 5-6B. Portion of the pipe section removed to reveal the interior of the pipe.

Notice in the sectional view in Fig. 5-5 that sectional lining or cross-hatching is drawn across those surfaces that are in contact with the cutting plane. This section lining is used in sectional views to indicate the various materials of construction. For the common metal—such as the galvanized steel of the conduit—lining is made with fine lines, usually drawn at angles of 45 degrees. A few of the common section-lining symbols are given in Fig. 5-8.

Sectional views of other construction equipment are classified as:

1. full section
2. half section
3. offset section
4. revolved section
5. removed section
6. aligned section
7. broken-out section
8. assembly section

Fig. 5-7. View of pipe with cutting plane removed.

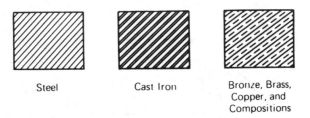

Steel   Cast Iron   Bronze, Brass,
                              Copper, and
                              Compositions

Fig. 5-8. Some common section-lining symbols.

Full section: A full section is a sectional view in which the cutting plane is assumed to pass entirely through the object. The sectional view in Fig. 5-9 is called a full section.

Half section: A half section is a sectional view in which the cutting plane passes halfway through the object. One half of the view is shown in section, while the other half is shown from the exterior. Figure 5-10 shows a cutting plane passing halfway through a piece of underfloor round duct which is constructed of tile. Figure 5-10 shows a portion in front of the cutting plane removed, and this portion is indicated by the cross-hatched lines.

Offset section: An offset section is a sectional view in which the cutting plane is bent, or offset, as shown in Fig. 5-11.

Revolved section: A revolved section is a cross section that has been revolved through 90 degrees. It is used to show the true shape of the cross section of bars and other elongated parts. Figure 5-12 shows an example of such a section.

Removed section: A removed, or detail, section is a cross section that has been removed from its original position to a convenient space near the principal view, such as the section shown in Fig. 5-13.

Aligned section: An aligned section is a sectional view in which a sloping part is rotated parallel to the cutting plane to show its true shape, such as the section shown in Fig. 5-14.

Broken-out section: A broken-out section, as shown in Fig. 5-15, is used when less than a half section is sufficient to show some interior detail. An irregular break line separates the section from the exterior view.

Assembly section: Assembly drawings are often drawn in section to show how the interior parts are fitted together. However, parts that lie

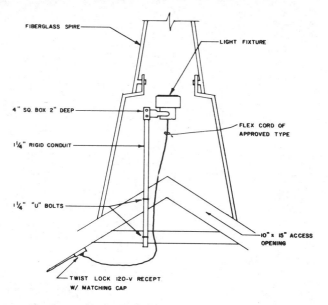

Fig. 5-9. Full section.

Fig. 5-10. Half section.

(a) Exterior View

Fig. 5-11. Offset section.

(b) Sectional View

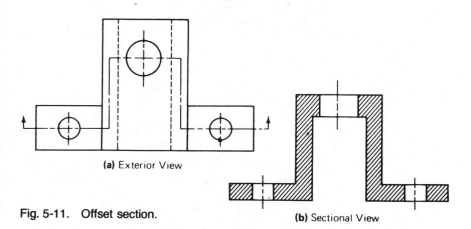

Fig. 5-12. Revolved section.

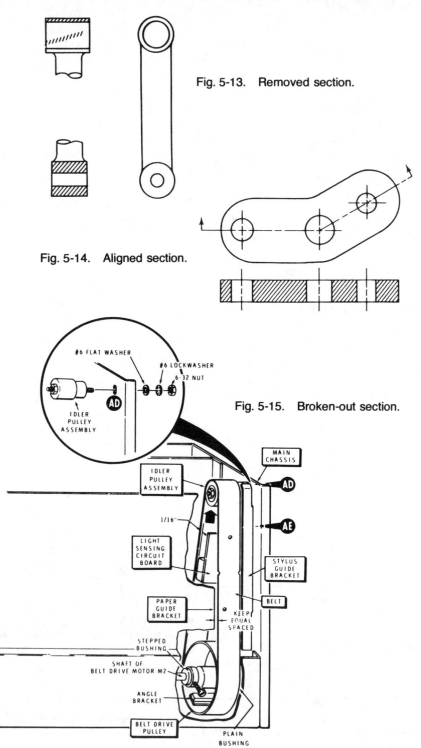

Fig. 5-13.   Removed section.

Fig. 5-14.   Aligned section.

Fig. 5-15.   Broken-out section.

#6 FLAT WASHER

#6 LOCKWASHER
6-32 NUT

AD

IDLER
PULLEY
ASSEMBLY

MAIN
CHASSIS

IDLER
PULLEY
ASSEMBLY

AD

1/16"

AE

LIGHT
SENSING
CIRCUIT
BOARD

STYLUS
GUIDE
BRACKET

PAPER
GUIDE
BRACKET

BELT

KEEP
EQUAL
SPACED

STEPPED
BUSHING

SHAFT OF
BELT DRIVE MOTOR M2

ANGLE
BRACKET

BELT DRIVE
PULLEY

PLAIN
BUSHING

in the path of the cutting plane, such as bolts, screws, etc., are not drawn in section. See Fig. 5-16.

Hidden Lines

Since a section is cut through an object to show the interior, all hidden lines are normally omitted; however, hidden lines do sometimes appear on a drawing in views where some hidden objects or line are absolutely indispensable for clarification or for dimensioning. In half sections, for example, hidden lines should be used only on the unsectioned side, and when they are absolutely necessary for dimensioning or clarity. See Fig. 5-17.

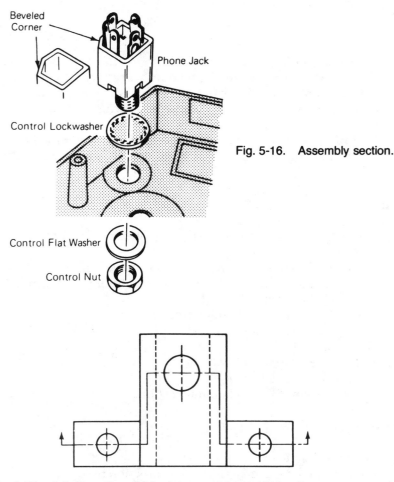

Fig. 5-16.   Assembly section.

Fig. 5-17.   A half section which also necessitates hidden lines.

Conventional Breaks

Because of their required size, the entire length of some elongated objects may sometimes be impossible or inconvenient to draw. The practice in these cases is to shorten the view by using break lines. Figure 5-18 shows how break lines are used to shorten elongated objects.

Fig. 5-18.   Break lines used to shorten elongated objects.

Summary

A section of any object is what could be seen if the object were sliced or sawed open at the point where the section is taken. Without sections, hidden (dotted) lines would have to be used to show all invisible details and these are seldom clear, causing the drawing to be difficult to read.

Section lining is usually shown across those surfaces which are in contact with the cutting plane. All visible details behind the cutting plane must also be shown.

Questions

1. For what purpose are cutting line planes used? How are they drawn?
2. Name one purpose for which section lining is used.
3. Draw a rectangle and then draw section lining of cast iron.
4. What is a full section?
5. Write the definition of a half section.
6. Define an offset section.
7. Draw the section lining for brass.
8. Define a revolved section.
9. What is a removed section?
10. What is meant by a broken-out section?

Answers to Questions

1. To show where a section is taken.
2. To indicate various materials of construction in sectional views.
3. See illustrations in text.

4. A sectional view in which the cutting plane is assumed to pass entirely through the object.
5. A sectional view in which the cutting plane passes only half-way through the object.
6. A sectional view in which the cutting is bent or offset.
7. See illustrations in text.
8. A cross section that has been revolved through 90°.
9. A cross section that has been removed from its original position.
10. A section that is used when less than a half section is sufficient to show a particular interior detail.

6 Diagrams

Objectives

*To introduce construction drawing diagrams so that the
reader will become familiar with this type of drawing.*

The building construction worker will encounter various types of
diagrammatic drawings more often on electrical and mechanical working drawings than any other types. However, those persons required to
interpret construction drawings should have a thorough understanding
of all types of drawings in order to approach his work more intelligently.
Electrical wiring diagrams include the following:

1. Diagrammatic plan views showing individual building-circuit
layouts. See Fig. 6-1.

2. Complete schematic diagrams showing all details of connection
and every wire in the circuit as shown in Fig. 6-2.

3. Single-line diagrams. See Fig. 6-3.

4. Power-riser diagrams.

Mechanical diagrams include the following:

1. Diagrammatic plan views showing water supply piping, drains,
and so forth. See Fig. 6-4.

2. Diagrammatic drawings of air ducts, grilles, diffusers, etc. as
shown in Fig. 6-5.

3. Riser diagrams showing drainage systems. See Fig. 6-6.

4. Refrigeration flow diagrams as shown in Fig. 6-7.

5. Schematic wiring diagrams showing the connection of control
circuits. See Fig. 6-8.

6. Schematic diagrams showing the components and connections
of pneumatic controls.

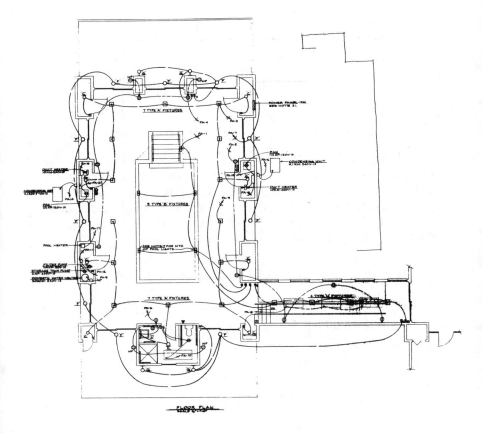

Fig. 6-1.   Plan view of building showing electrical circuits.

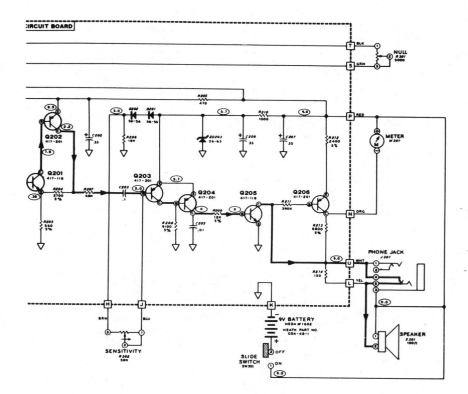

Fig. 6-2. Schematic diagram showing all details of the circuit.

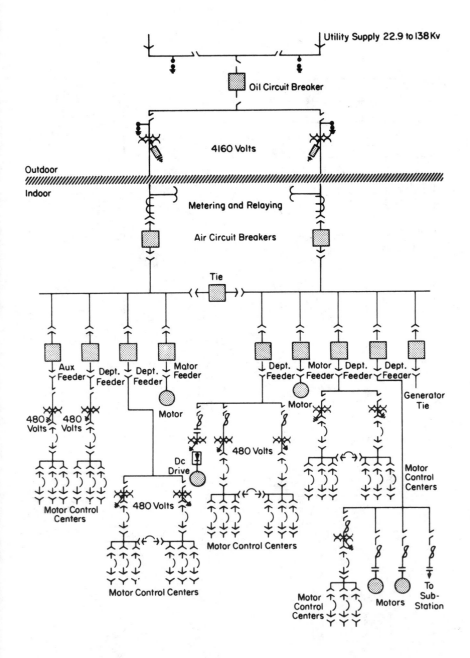

Fig. 6-3.   Single-line diagram.

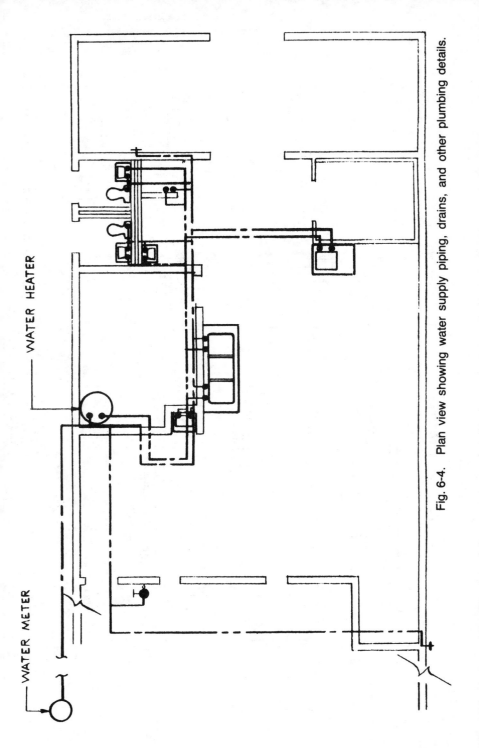

WATER HEATER

WATER METER

Fig. 6-4. Plan view showing water supply piping, drains, and other plumbing details.

87

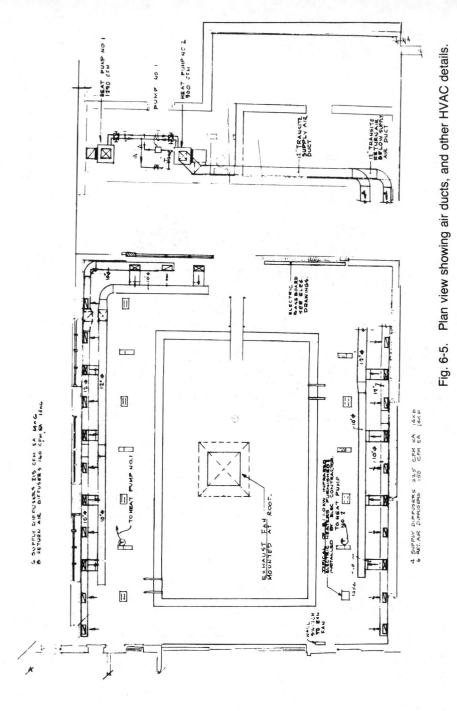

Fig. 6-5. Plan view showing air ducts, and other HVAC details.

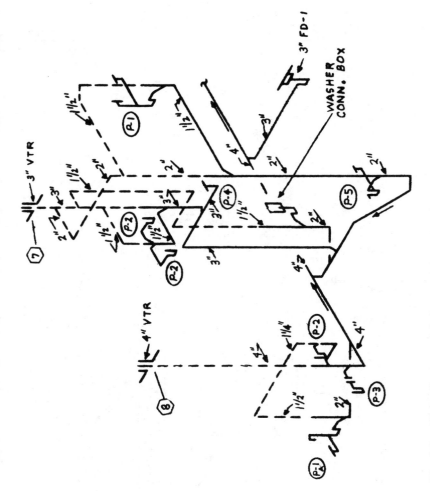

Fig. 6-6. Riser diagram of drainage system.

89

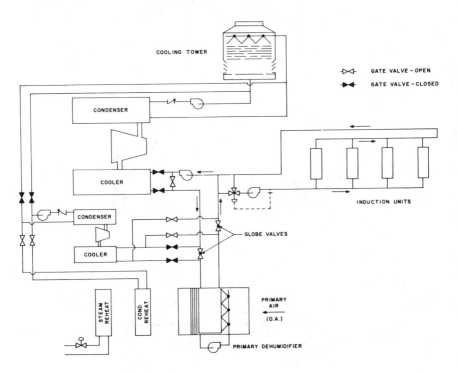

Fig. 6-7.   Refrigeration flow diagram.

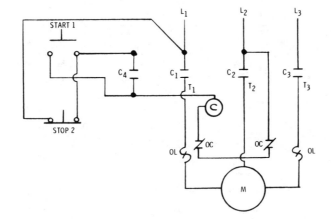

Fig. 6-8.   Schematic
wiring diagram of a con-
trol circuit.

On complicated electrical systems, such as highly sophisticated control circuits, complete schematic wiring diagrams are usually provided in the working drawings. The various components such as relays, switches, and the like are represented by symbols, and every wire or conductor is either shown individually or included in an assembly of several wires which appear as one line on the drawing. In the latter case, each wire is usually numbered when it enters an assembly and keeps the same number when it emerges from the assembly to be connected to some electrical component in the system.

Power-riser diagrams such as the one shown in Fig. 6-9 are probably the most frequently used type of diagrams on electrical working drawings for building construction. Such diagrams give a picture of what components are to be used and how they are to be connected in relation to one another. This type of diagram is easily understood and requires much less time to interpret than schematic diagrams. As an example, compare the power-riser diagram in Fig. 6-10 with the schematic diagram in Fig. 6-11. Both are diagrams of an identical electrical system, but it is easy to see that the drawing in Fig. 6-10 is greatly simplified.

Plumbing Diagrams

Methods of showing plumbing layouts on working drawings will vary with each consulting engineering firm or contractor, but the following description is typical for most firms.

All domestic cold water piping is normally laid out on the building floor plans as shown in Fig. 6-11. Note that all valves, stops, hose bibbs, and other connections are indicated by symbols and the water lines themselves are indicated by broken lines; the ---- pattern on the drawing indicates a cold water line and the --·-- pattern indicates a hot water line. However, always check the legend or symbol list on each set of drawings, as these symbols could be reversed or altogether different.

Drains and vents in the drawing in Fig. 6-12 are shown by symbol only; no piping is shown in this drawing. When such a case exists, a piping riser diagram such as the one in Fig. 6-13 is normally used to diagrammatically show the arrangement of the drain and vent pipes. The drawing in Fig. 6-13 is an orthographic view, but such diagrams are sometimes shown by means of an isometric drawing (See Fig. 6-14).

Pipe sizes and other pertinent details are usually indicated by numerals or notes respectively and appear immediately adjacent to the component or equipment described. Schedules and written specifications complete the set of plumbing working drawings. The schedules include such items as the manufacturer, catalog number, pipe connections, etc. of all plumbing fixtures and similar equipment.

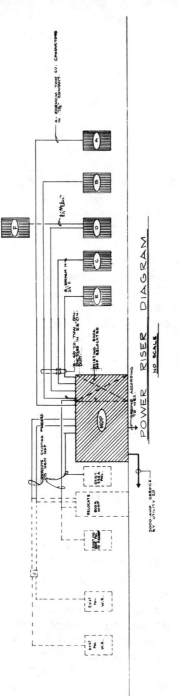

Fig. 6-9. Power riser diagram.

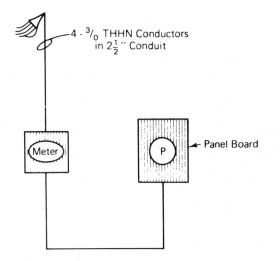

Fig. 6-10.   Simplified drawing of a power riser diagram.

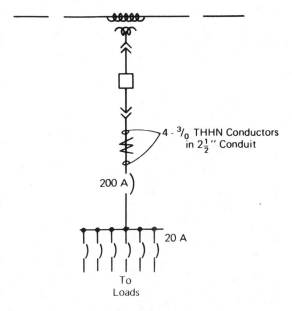

Fig. 6-11A.   Schematic diagram of an electric service.

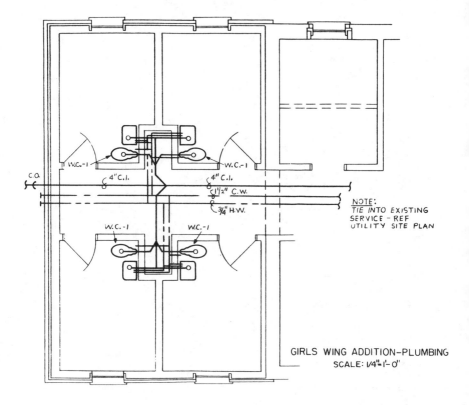

Fig. 6-11B.   Building floor plan showing plumbing layout.

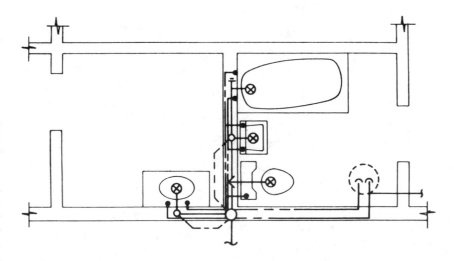

Fig. 6-12.   Plumbing drains and vent shown on drawing in symbol form only.

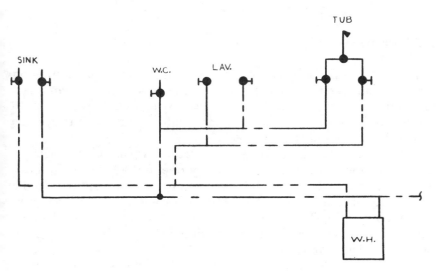

Fig. 6-13.   Piping riser diagram for the drawing in Fig. 6-12.

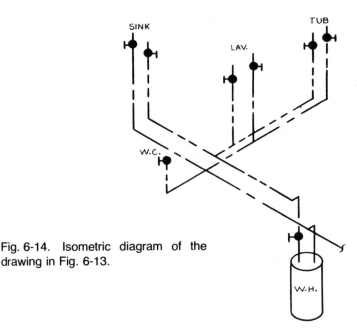

Fig. 6-14. Isometric diagram of the drawing in Fig. 6-13.

HVAC Drawings

The term HVAC stands for heating, ventilating, and air-conditioning and is a common term on most construction sites as well as in architectural and engineering firms. Some of the drawings are drawn similar to architectural drawings while others are highly diagrammatic such as electrical and plumbing drawings.

If outside oil and gas tanks are involved on a particular project, or natural gas lines must cross the property line to the building for heat, the HVAC drawings usually include a plot plan showing the details and routing of such lines and location of the equipment. Such equipment as air handling units, condensers, furnaces, etc. are located on the drawings in their proper location and drawn to scale in their approximate shape. Ductwork can be shown in two different ways: as scaled plan drawings as shown in Fig. 6-15, or as single-line drawings as shown in Fig. 6-16.

All air grilles, diffusers, and the like are located in their respective place on the floor plans and each is given a number to indicate the size and type of each. A schedule is usually provided to give further descriptions of the various air outlets and other pieces of equipment.

Controls circuits like the one shown in Fig. 6-17 should accompany all but the simplest control systems so that the system will function properly. When such diagrams are omitted, workmen waste much time in obtaining necessary information so that the controls can be connected.

Schematic Diagrams

Electronic controls for heating and cooling systems, security/fire alarm systems, intercoms and similar systems have made it necessary to use schematic diagrams in showing how these systems are installed and operated.

A schematic diagram is a graphic presentation that shows the connection and operational characteristics of an electronic circuit by means of lines, symbols, and notations. Besides showing how a system is to be installed, schematic diagrams are also invaluable for use in troubleshooting the system.

In general, electronic schematic diagrams indicate the scheme, or plan, according to which electronic components are connected for a specific purpose. Diagrams are not normally drawn to scale, and the symbol rarely looks exactly like the component. Lines joining the symbols representing electronic components indicate that the components are connected.

In order to serve satisfactorily all its intended purposes, the schematic diagram must be accurate, it must be understood by all qualified personnel, and it must provide definite information, without ambiguity.

The schematics for an electronic device must indicate all circuits in the device. Furthermore, it should be easy to read and follow an entire closed path in each circuit. If there are interconnections, they should be clearly indicated.

If at all possible, the conductors connecting the electronic symbols should be drawn either horizontally or vertically; rarely are they ever slanted.

A dot at the junction of two crossing wires means a connection between the two wires, while an absence of a dot—in most cases—indicates that the wires cross without connecting.

Schematic diagrams are, in effect, shorthand explanations of the manner in which an electronic circuit or group of circuits operates. As such, they make extensive use of symbols and abbreviations. The more commonly used symbols were explained earlier, and the reader should become thoroughly familiar with all of these symbols in order to correctly read electronic schematic drawings. The use of symbols presumes that the person looking at the diagram is reasonably familiar with the operation of the device and that he will be able to assign the correct meaning to the symbols.

Usually, every component on a complete schematic diagram has a

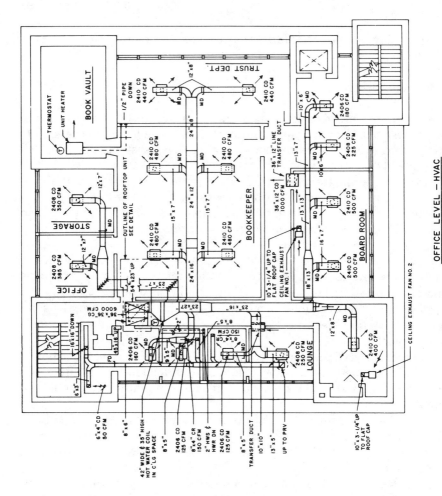

Fig. 6-15. Scale plan drawing of an HVAC system.

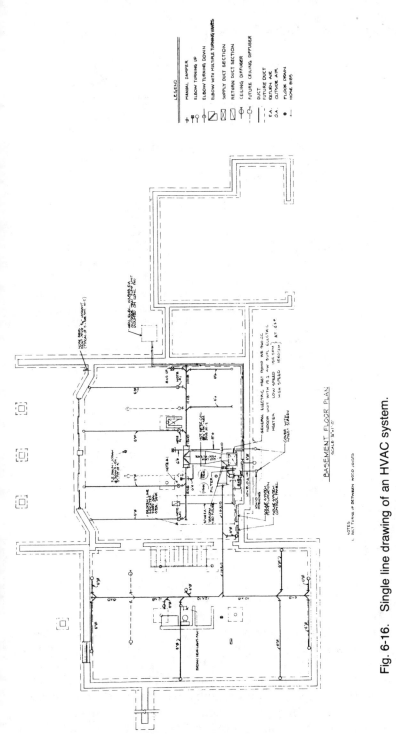

Fig. 6-16. Single line drawing of an HVAC system.

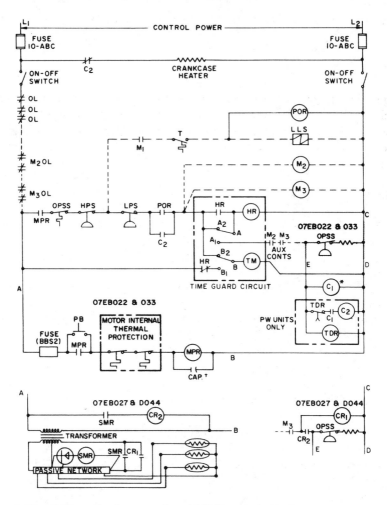

Fig. 6-17. Schematic diagram of an HVAC control circuit.

number to identify the component. Supplementary data about such parts are supplied on the diagram or on an accompanying list in the form of a schedule which describes the component in detail.

To interpret schematic diagrams, remember that each circuit must be closed in itself, and each component should be in a closed loop connected by conductors to a source of electric current, such as battery, transformer, etc. There will always be a conducting path leading from the source to the component and a return path leading from the component to the source. The path may consist of one or more conductors. There may also be other components in the same loop, or additional loops branching off to the other devices. But for each electronic component it must be possible to trace a completed conducting loop to the source.

Summary

In general, diagrams are graphic presentations of a system or systems that show the major components and how they are interconnected. Seldom are diagrams drawn to scale; rather, as the name implies, they are diagrammatic—showing only the components in relation to each other and their function.

Questions

1. Name the three building trades that use diagrams more so than others.
2. Draw the symbol on plumbing diagrams that normally indicate a hot water line.
3. What does the term HVAC stand for in the building construction industry?
4. What type of diagram is normally used to show control circuits for heating and cooling systems?
5. In an electronic schematic diagram, what does a dot at the junction of two crossing wires mean?

Answers to Questions

1. Electrical, plumbing, and mechanical
2. ——-——
3. Heating, ventilating, and air conditioning
4. schematic
5. a connection

7 Schedules and Their Use

Objectives

To acquaint the reader with schedules that appear on construction working drawings and to teach the reader how to read schedules of all types.

In general, a schedule, as applied to construction working drawings—is a systematic method of presenting notes or lists of materials, equipment, components, and the like on a drawing in tabular form. When properly organized and thoroughly understood, not only are they great time-saving devices for those preparing the drawings, but they also save the architect, the contractor, and the workmen on the job much valuable time.

The door schedule in Fig. 7-1, for example, lists the door type and identifies each door type on the drawings for a given project by number. The manufacturer and identification number of each type are given, along with the number, size, and type of each.

If this same data were to be specified in the written specifications, not only would it take the designer longer to describe the various doors and how to identify them on the drawings, but it would also require that the workmen on the job comb through page after page of the written specifications to obtain the same information. Many times, workmen do not always have access to the written specifications on the job, whereas they almost always have one or more copies of the working drawings.

From the previous paragraph, we can see that the schedule offers an excellent means of providing essential information in a clear and accurate manner, allowing the workmen to carry out their assignments in the least amount of time.

The following schedules are typical of those used on building construction drawings; each should be thoroughly studied along with the accompanying text.

DOOR SCHEDULE

MARK	TYPE	SIZE WIDE	SIZE HIGH	THK.	FRAME TYPE	LINTEL	T'HOLD	REMARKS	HOW HUNG N.S.
201	B	2'-10"	7'-0"	1¾"	DET. "D-6"	WOOD	---		3
202	B	2'-8'	7'-0"	1¾"	DET. "D-6"	WOOD	---		9
203	B	2'-10"	7'-0"	1¾"	DET. "D-6"	WOOD	---		9
204	B EXIST.	3'-0"	7'-0"	1¾"	DET. "D-6"	WOOD	---	REUSE DOOR FROM CORR. TO SPACE 200	3
205	B EXIST.	3'-0"	7'-0"	1¾"	DET. "D-6"	WOOD	---	REUSE DOOR FROM CORR. TO SPACE 205	6
206	B	2'-8"	7'-0"	1¾"	DET. "B-6"	EXIST.	---		5
207	B	2'-8"	7'-0"	1¾"	DET. "B-6"	EXIST.	---		3
208	B	2'-6"	7'-0"	1¾"	DET. "D-6"	WOOD	CARPET STRIP		11
209	B	2'-6"	7'-0"	1¾"	DET. "D-6"	WOOD	CARPET STRIP		3
210	B	2'-8"	7'-0"	1¾"	DET. "D-6"	WOOD	---		4
211	B	2'-8"	7'-0"	1¾"	DET. "D-6"	WOOD	---	NOTE 12"x16" LOUVRE FURNISHED BY MECH. (SEE MHT. M-2) FOR INSTALLATION BY CONTRACTOR.	3
212	B	2'-8"	7'-0"	1¾"	DET. "D-6"	WOOD	CARPET STRIP	SEE NOTE FOR DOOR NE 211 ABOVE	3
213	B	2'-8"	7'-0"	1¾"	DET. "D-6"	WOOD	CARPET STRIP		3
214	B	2'-8"	7'-0"	1¾"	DET. "D-6"	WOOD	CARPET STRIP		3
215	D	2'-8"	7'-0"	1¾"	DET. "E-6"	WOOD	---		3
216	D	2'-8"	7'-0"	1¾"	DET. "C-6"	EXIST.	---		10

Fig. 7-1. Typical door schedule.

Material Schedules

The material schedule in Fig. 7-2 lists the approximate quantities of materials needed to complete the project in question. Such schedules are used mainly for estimating purposes on small construction projects, such as residential buildings. Unless specifically stated otherwise in the specifications, these material schedules should never be taken as the exact amount of material needed. They are used merely as a check or guide for experienced estimators.

Even on repetitive projects, where accurate records have been kept, the exact amount of materials varies so much that it is very unlikely that a highly accurate schedule could ever be compiled.

Lighting Fixture Schedules

The lighting-fixture schedule in Fig. 7-3 lists the fixture type and identifies each fixture type on the drawings for a given project by number. The manufacturer and identification number of each type are given, along with the number, size, and type of the lamps for each. The "Volts" column gives the correct voltage to which the fixture is to be connected, and the "Mounting" column tells whether the fixture is wall-mounted, surface-mounted on the ceiling, or recessed. The remaining column may give such information as the mounting height above the finished floor, in the case of a wall-mounted lighting fixture, or any other pertinent data for the proper installation of the fixtures.

Panelboard Schedules

The panelboard schedule in Fig. 7-4 is typical of those used on electrical working drawings to give pertinent data on the service panel boards within a building. This schedule provides sufficient data to identify the panel number (as shown on the drawings), the type of cabinet (either surface-mounted or flush), the panel main bus bars and/or circuit breaker (in amperes, volts, and phase), the number and type of circuit breakers contained in the panel board, and the items fed by each. This type of schedule, however, does not give detailed information concerning the individual circuits, such as the wire size or the number of outlets on the circuit; this latter information must be given elsewhere on the drawing—usually in the plan view or in power-riser diagrams.

HVAC Schedules

The following schedules appear on mechanical drawings for heating, ventilating, and air-conditioning systems.

LINE NO.	ITEM N° 1080 COLUMN NO. 1	QUANTITY & UNIT MEAS.	MATERIAL (TYPE and/or SIZE)	UNIT COST	TOTAL COST	LINE NO.	ITEM N° 1080 COLUMN NO. 2	QUANTITY & UNIT MEAS.	MATERIAL (TYPE and/or SIZE)	UNIT COST
1	MASONRY					1	FRAMING LUMBER (Continued)			
2	Footings	84 Cu.Yds.	60/40 Gravel			2	Exterior Walls	490 Pcs.	2 x 4 x 8'-0" Studs	
3	96 Lin.Ft. 80x10	336 Sacks	Cement			3		1300 Lin.Ft.	2 x 4 Plates	
4	184 Lin.Ft. 16x8	158 Pcs.	4" Drain Tile			4		84 Pcs.	4"x 4"x 7'-0" Posts (Clear ladder stock) fir	
5	300 Lin.Ft. 8x42	10 Pcs.	4" " Els			5		808 Lin.Ft.	2 x 6 Headers	
6	850 Lin.Ft.12x42	6 Pcs.	4" " Tees			6	8 Beams	3- 2x 12x 18'-0" with 3/8"x 11"x 16'-0"	Steel Pla
7	1 Pad 1'-dx 7'-dx	6 Pcs.	4" Vitrified Crock			7			Garage Beams	
8	18	3 Rolls	15# 4" Strip Felt			8	8 Beams	3- 2x 12x 16'-0" with 3/8"x 11"x 15'-8"	Steel Pla
9	1 Pad 4'-0x 18'-0x	10 Cu.Yds.	Pea Gravel			9			Breezeway Beams	
10	18	500 Sq.Ft.	1" Fiberglass Perimeter Insulation			10	4 Pcs.	2 x 8 x 8'-0" Headers	
11	4 Pads 24x24x18					11	2 Pcs.	2 x 10 x 16'-0" "	
12						12	4 Pcs.	2 x 10 x 12'-0" "	
13	Masonry Block Walls					13	2 Pcs.	2 x 10 x 12'-0" "	
14		70 Pcs.	12 x 8 x 16" Grade Masonry Blocks			14	2 Pcs.	2 x 10 x 24'-0" "	
15		8 Pcs.	12 x 8 x 16" Corner " "			15	2500 Sq.Ft.	15# Felt	Wall Sheathing
16		630 Pcs.	12 x 8 x 16" Regular " "			16	3600 Sq.Ft.	15# Felt	
17		794 Pcs.	8 x 8 x 16" " "			17	Interior Partition			
18		54 Pcs.	8 x 8 x 16" Corner " "			18		80 Pcs.	2 x 4 x 12'-0" Studs	
19		126 Pcs.	8 x 8 x 16" Solid " "			19		15 Pcs.	2 x 4 x 10'-0" "	
20		164 Pcs.	8 x 8 x 16" " Masonry "L" Blocks			20		380 Pcs.	2 x 4 x 8'-0" "	
21		60 Pcs.	4 x 8 x 12" " " Blocks			21		85 Pcs.	2 x 6 x 8'-0" "	
22		94 Pcs.	4 x 8 x 16" " "			22		1300 Lin.Ft.	2 x 4 Plates	
23		10 Cu.Yds.	50/50 Mason Sand			23		80 Lin.Ft.	2 x 8 "	
24		50 Sacks	Mortar			24		1 Unit	2'-0"x 6'-8"x 3-3/8" Sliding Door Pocket	& Track
25		15 Sacks	Cement			25		2 Units	3'-0"x 6'-8"x 3-3/8" " " " " "	
26		20 Gals.	Asphalt Foundation Coating			26		330 Lin.Ft.	2 x 6 Header	
27		2 Pcs.	8 x 8 Cleanout Doors			27		64 Lin.Ft.	2 x 8 "	
28		1 Pc.	8" Diam. Steel Furnace Thimble			28	Ceiling Framing	4 Pcs.	2 x 12 x 30'-0" Wood Beams	
29		1 Pc.	6" Diam. Steel Water Heater Thimble			29		1 Pc.	3-1/4"x 16-1/4"x 24'-0" Structural Laminated Wood	
30		4 Pcs.	12 x 12 Flue Lining			30		1 Pc.	3 x 12 x 24'-0" Valance Drop	
31	Basement Steel &					31				
32	Sash	2 Pcs.	8 x 8 x 1/4" Steel Bearing Plates			32	Ceiling Framing	14 Pcs.	2 x 8 x 12'-0" Ceiling Joist	
33		4 Pcs.	3½" Diam. Steel Pipe Columns			33		12 Pcs.	2 x 6 x 14'-0" "	
34		1 Pc.	8" 10.4# 42'-1" Steel "I" Beam			34		48 Pcs.	2 x 6 x 16'-0" "	
35		7 Pcs.	3-Light 15 x 12" Steel Sash			35		16 Pcs.	2 x 6 x 20'-0" "	
36		7 Pcs.	36" Diam. x 24" Corrugated Steel Areaways			36		80 Pcs.	2 x 6 x 24'-0" "	
37	Fireplace &					37	Roof & Cornice			
38	Veneer	6 Pcs.	3"x 3½"x 3/8" Fireplace Lintel Angles			38	Framing	160 Lin.Ft.	2 x 8 Ridge Member	
39		2 Pcs.	48" Dome Damper			39		7 Pcs.	2 x 8 x 22'-0" Hip Member	
40		240 Pcs.	Firebrick			40		2 Pcs.	2 x 8 x 18'-0" "	
41		75 Pcs.	Fireclay			41		1 Pc.	2 x 8 x 16'-0" "	
42		1 Pc.	2-1/4"x 8"x 4'-6" Flagstone Hearth			42		48 Pcs.	2 x 6 x 12'-0" Rafters	
43		1 Pc.	2-1/4"x 8"x 6'-8" " Mantel Surround			43		34 Pcs.	2 x 6 x 14'-0" "	
44		2 Pcs.	2-1/4"x 8"x 2'-6" "			44		170 Pcs.	2 x 6 x 16'-0" "	
45		1 Pc.	5 x 8 Ash Dump			45		74 Pcs.	2 x 6 x 18'-0" "	
46		1 Unit	36"x 88" Double Barbeque			46		464 Lin.Ft.	2 x 6 Rafter Trimmers	
47		1 Pc.	18"x 28"x2-1/4" Flagstone Barbeque Counter			47		1370 Lin.Ft.	2 x 4 Cornice Soffit Framing	
48		9 Pcs.	12 x 16" Flue Lining			48		6844 Sq.Ft.	Roof Sheathing	
49		7 Pcs.	12 x 12" "			49		14000 Sq.Ft.	15# Felt	
50		300 Pcs.	4 x 8 x 16" Regular Masonry Blocks			50				
51			(Plaster Linings)			51	ROOFING & SHEET METAL			
52		7 Sacks	Cement "			52		3 Squares	Asphalt Ridge & Hip Shingles	
53		2 Sacks	Asphalt Foundation Coating (Plaster Lining)			53		68 Squares	Asphalt Self Sealing "	
54		36 Pcs.	1-1/2"x 3-3/4"x 14" Roman Brick (Mantel Lining)			54		250 Lbs.	Roofing Nails	
55		7000 Pcs.	Common Brick (Chimneys)			55		100 Pcs.	5 x 7 Metal Step Flashing	
56		172 Lin.Ft.	2-1/4"x 5" Flagstone Sill			56		52 Lin.Ft.	Chimney Counter Flashing	
57		160 Lin.Ft.	2-1/4"x 9" " Coping			57		490 Lin.Ft.	Metal Drip Edging	
58		9 Lin.Ft.	2-1/4"x 13" "			58				
59		16 Lin.Ft.	2-1/4"x 5" "			59	WINDOW			
60		2612 Sq.Ft.	4" Cut Stone Veneer			60		22 Single	12 x 80 Fixed Plate Glass - Loose Casing	
61		35 Cu.Yds.	50/50 Mason Sand			61		4 Single	90 x 80 " "	
62		200 Sacks	Mortar			62		1 Single	82 x 80 " "	
63		1300 Pcs.	Metal Wall Ties			63		1 Single	68 x 80 " "	
64		1 Pc.	4 x 11 x 6'-0" Cut Stone Door Sill			64		3 Single	44 x 80 " "	
65		1 Pc.	4 x 11 x 4'-0" "			65		3 Single	24 x 36 Casements - Brick Mold Casing	
66						66		4 Triple	24 x 36 " " " "	
67	CONCRETE SLABS					67		2 Triple	24 x 36 " " " "	
68	Garage & Terraces	(Sidewalks, Driveways and pool not included)				68		1 Single	30 x 18 Awning " " "	
69		3000 Sq.Ft.	6 x 6 #6/6 Welded Wire Reinforcing Mesh			69		3 Single	42 x 18 " " " "	
70		54 Cu.Yds.	Fill Gravel			70				
71		42 Cu.Yds.	60/40 Gravel			71	DOOR FRAMES			
72		210 Sacks	Cement "			72		1 Front	5'-4"x 6'-8" Rabbeted 1-3/4" Loose Casing	
73		50 Gals.	Liquid Cement Hardner			73		1 Rear	3'-0"x 6'-8" " 1-3/4" Brick Mold Casing	
74	House Floors					74		1 Garage	9'-0"x 7'-0" 2x6 Overhead Frame Brick Mold Casing	
75		2300 Sq.Ft.	Plastic Membrane Vapor Barrier			75		1 Rear	2'-8"x 6'-8" Rabbeted 1-3/4" Brick Mold Casing	
76		46 Cu.Yds.	Fill Gravel			76		1 Terrace	12'-0"x 7'-0" " Glass Sliding Door, Frame	
77		46 Cu.Yds.	60/40 Gravel			77			and Hardware	
78		230 Sacks	Cement			78		3 Terrace	8'-0"x 7'-0" " Glass Sliding Door, Frames	
79		48 Pcs.	4" Vitrified Crock			79			and Hardware	
80		5 Pcs.	4" " 45° Bends			80				
81		5 Pcs.	4" " Tees			81	Loose Door &			
82						82	Window Casing	264 Lin.Ft.	1-1/8"x 1-3/4" Brick Mold Casing	
83	FRAMING LUMBER					83		390 Lin.Ft.	1-1/2"x 4" Mullion Facing	
84	Platform	160 Lin.Ft.	2 x 6 Sill Plate			84		1 Pc.	1-3/8"x 10"x 16'-0" Garage Door Head Casing	
85		96 Lin.Ft.	2 x 10 Joist Trimmers			85		24 Lin.Ft.	1-3/8"x 6" Garage Window Head Casing	
86		15 Pcs.	2 x 10 Joist "			86				
87		32 Pcs.	2 x 10 x 12'-0" "			87	EXTERIOR TRIM			
88		10 Pcs.	2 x 10 x 14'-0" "			88	Cornice	464 Lin.Ft.	1 x 6 Fascia	
89		50 Pcs.	2 x 10 x 16'-0" "			89		490 Lin.Ft.	1 x 6 Slotted & Screened Cornice Vent Board	
90		184 Lin.Ft.	1 x 8 Bridging			90		74 Pcs.	4 x 8 x 1/4" Exterior Grade Plywood Soffit &	
91		1100 Sq.Ft.	1/2" Plywood Subfloor			91			Porch Ceiling	

Fig. 7-2. Material schedule listing the approximate quantities of materials for a given project.

LIGHT FIXTURE SCHEDULE

FIXT TYPE	MANUFACTURER DESCRIPTION	LAMPS No.	LAMPS TYPE	VOLTS	MOUNTING	REMARKS
A	LIGHTCRAFT No.-S-IH	1	150 WI	120	RECESSED	POOL AREA LIGHTING
B	HUBBELL No.-414	1	116 WI	120	RECESSED	UNDERWATER LIGHTS
C	LIGHTCRAFT No.-802	1	60 WI	120	SURFACE	STORAGE ROOMS.
D	LIGHTCRAFT No.-163-72-76*	2	60 WI	120	WALL	RESTROOM
E	ART METAL No.-3357AA	1	100 WI	120	SEMI REC.	RESTROOM
F	LIGHTCRAFT No.-96	1	100 WI	120	RECESSED	
G	VIRDEN CAMBRIDGE No.-V-7160	1	100 WI	120	WALL	PATIO
H	ART METAL No.-15-15TL-WS	1	150 WI	120	RECESSED	CORRIDOR.

Fig. 7-3. Typical lighting fixture schedule.

PANELBOARD SCHEDULE

PANEL Nº	TYPE CABINET	MAINS AMPS	MAINS VOLTS	PHASE	1P	2P	3P	PROT	FRAME	ITEMS FED
A	SURFACE	400 A	120/208V	3Ø,4W	30	-	-	20A	100	LTG & RECEPTS.
					12	-	-	-	-	PROVISIONS ONLY
SQ."D" OR EQUAL (NQO) W/ M.L.O.										
B	SURFACE	225A	120/208V	3Ø,4W	30	-	-	20A	100	LIGHTS & RECEPTS.
					12	-	-	-	-	PROVISIONS ONLY
SQ."D" OR EQUAL (NQO) W/ M.L.O.										
C	SURFACE	400A	120/208V	3Ø,4W	-	-	2	100A	100	CONDENSING UNITS
					-	-	2	30A	-	FAN COIL UNITS
					12	1	1	-	-	PROVISIONS ONLY
SQ."D" OR EQUAL (NQO) W/ M.L.O.										

Fig. 7-4. Electrical panelboard schedule.

Grille Schedule

A grille and diffuser schedule (Fig. 7-5) shows the manufacturer and catalog number of each grille and diffuser; it also shows the dimensions of each and the volume of air in cubic feet per minute (CFM) that each will handle. A column for remarks is also included to facilitate the installation of the item.

Hot-Water-Boiler Schedule

Figure 7-6 shows an example of a hot-water-boiler schedule used on an actual construction project.

If this schedule were not indicated on the drawings, a description of the boiler would have to appear in the written specifications and would be similar to the following:

Contractor shall furnish and install, as indicated on plans, one oil-fired boiler having a firing rate of 2.55 gallons per hour (gph) and an I.B.R. gross output of 261,000 Btuh. Boiler shall be constructed of cast iron in accordance with ASME (American Society of Mechanical Engineers) requirements for low-pressure heating boilers and bear the ASME symbol. Each section shall be factory tested at 2½ times maximum working pressure of 50 pounds for water. Boiler shall have I-B-R rating.

Boiler shall be composed of cast iron sections with vertical flues. Manufacturers will furnish boilers with an aluminized steel canopy. Water or steam trim as required. Oil burners shall include controls as herein described.

Boiler shall be equipped with tankless heater having a rating of 7 gpm. If more than one heater is required to meet requirements, a factory-assembled copper manifold will be installed by the contractor as recommended by the manufacturer to connect the heater in parallel arrangement.

Diffuser Schedule

Figure 7-7 shows another type of diffuser schedule used on the HVAC drawings for a large residence. The left-hand column indicates the room or area in which the diffusers are located; the next column, reading to the right, indicates the quantity, manufacturer's name, model number, and dimensions of the supply air diffusers for each area; the next column indicates the air volume in cubic feet per minute (CFM); and the last column is left empty for remarks.

Exhaust Fan Schedule

The exhaust fan schedule illustrated in Fig. 7-8 was used for the same residence previously discussed. The left-hand column gives the room or area designation, and the remaining column indicates the manufacturer, model number, cubic feet per minute, and the sound levels in

SUPPLY DIFFUSER SCHEDULE LOWER LEVEL

GUEST B.R.	2-CARNES MODEL 7292-A x 4'0" LONG	157-CFM EACH	
GAME RM.	BASE BOARD ELECTRIC		
BATH #1	1-CARNES MODEL 200-12x4 REG.	NO VOLUME CONTROL 50-CFM EACH	STACK HEAD DAMPER
BAR	1-CARNES MODEL 200-12x4 REG.	NO VOLUME CONTROL 121-CFM EACH	STACK HEAD DAMPER
KITCHENETTE	1-CARNES MODEL 200-12x4 REG.	NO VOLUME CONTROL 64-CFM EACH	STACK HEAD DAMPER
GALLERY	1-CARNES MODEL C-40 FLOOR 3"x5'0" / 2-CARNES MODEL 7261-A x 3'0" LONG	200-CFM EACH / 100-CFM EACH	
STUDY	1-CARNES MODEL 7292-A x 6'0" LONG	282-CFM EACH	
BED RM. #1	1-CARNES MODEL 7261-A x 6'0" LONG	188-CFM EACH	
BED RM. #2	1-CARNES MODEL 7261-A x 6'0" LONG	188-CFM EACH	
BATH #2	1-CARNES MODEL 200-12x4 REG.	NO VOLUME CONTROL 50-CFM EACH	STACK HEAD DAMPER
LAV. #1	1-CARNES MODEL 200-12x4 REG.	NO VOLUME CONTROL 50-CFM EACH	STACK HEAD DAMPER
LAV. #2	1-CARNES MODEL 200-12x4 REG.	NO VOLUME CONTROL 50-CFM EACH	STACK HEAD DAMPER

RETURN AIR DIFFUSER SCHEDULE

GAME RM.	BASE BOARD ELECTRIC		
GUEST B.R.	CARNES MODEL 7295 x 4'0"	250-CFM EACH	
GALLERY	CARNES MODEL 7295 x 4'0"	400-CFM EACH	
BED RM. #1	CARNES MODEL 7295 x 4'0"	238-CFM EACH	
BED RM. #2	CARNES MODEL 7295 x 4'0"	238-CFM EACH	
STUDY	CARNES MODEL 7295 x 4'0"	332-CFM EACH	

Fig. 7-5. A grille and diffuser schedule.

HOT WATER BOILER SCHEDULE							
BOILER NO.	FUEL	BTU/GAL.	BTUH-INPUT	OIL - GPH	IBR- NET	IBR-GROSS	FLUE
*A-504	Nº2 OIL	261 MBH	300,150	2.55	227,000	261,000	8"x 12"

*AMERICAN STANDARD

Fig. 7-6.   Example of a hot water boiler schedule.

GRILLE SCHEDULE					
MARK	MFGR	MODEL	SIZE	CFM	REMARKS
1	KRUEGER	SHI O	9 x 9	146	OBD ALUMINUM
2		SH 3	6 x 6	50	
3				60	
4				75	
5				80	
6				100	
7			9 x 9	144	
8				152	
9				170	
10				175	
11				181	
12				226	
13		SH 4	6 x 6	65	
14				100	
15			9 x 9	175	
16				222	
17		S580H	8 x 6	150	
*18			14 x 8	292	
19				300	
*20			18 x 8	354	
21			24 x 8	500	
22		S580V	34 x 12	1000	
23		SH 3	9 x 9	205	
24	KRUEGER	S580V	8 x 6	167	C.R. OBD ALUM
25		S580V	18 x 8	362	C.R. OBD ALUM

* INCLUDE FILTER FRAME   5FF HINGED

Fig. 7-7.   Diffuser schedule used on a residential HVAC drawing.

EXHAUST FAN SCHEDULE	
BATH No. I	NUTONE MODEL QT-110-110 CFM DECIBELS SOUND LEVEL 2.5
BATH No. 2	NUTONE MODEL QT-80-80 CFM DECIBELS SOUND LEVEL 1.5
LAY # I	" " " " " " " " "
LAY # 2	" " " " " " " " "
LAUNDRY RM.	2 EA NUTONE HEAT-A-VENT # 9275-4.265 B.TUH HEAT 40 CFM EA
	WITH 2 EA ON/OFF HEAT/VENT CONTROLS
MASTER BATH	I EA NUTONE HEAT-A-VENT # 9275-4.265 B.TUH-HEAT 40 CFM WITH
	ONE EA ON/OFF HEAT CONTROL WALL MOUNTED THERMOSTAT

Fig. 7-8.   Typical exhaust fan schedule.

decibels of the conventional fans. The heat output from the combination heat and fan units in the laundry room and master bath, as well as the controls, is listed in British thermal units per hour.

Radiation Schedule

Hot-water or steam baseboard radiators may be described in a schedule such as the one illustrated in Fig. 7-9. This schedule is explained as follows:

Symbol   R/1 as indicated in the schedule means that this line describes radiator No. 1, which is also identified in the same manner on the floor plans. However, the schedule describes the radiator in detail, while the floor plan gives the symbol R/1 only.

Serving   This particular schedule was used on a bank project. Note that the second column indicates that radiator No. 1 serves the Note Department, radiator No. 2 serves the vault and office area, etc.

Capacity MBH (thousand British thermal units per hour)   This column indicates the heat output, in British thermal units, of each unit. It can be seen that radiator No. 1 (R/1) has a capacity of 18.1 MBH, or an output of 18,000 Btuh.

Enclosure Length   This column gives the physical length of each radiator listed. This information is provided for two reasons: first, to indicate the required length in case the contractor submits a substitute

RADIATION SCHEDULE

SYMBOL	SERVING	CAPACITY MBH	ENCLOSURE LENGTH	ACTIVE LENGTH	RATING – BTU LIN. FT.	PROTO-TYPE	TYPE OF ENCLOSURE BACK	TYPE OF SLEEVE	REMARKS
R/1	NOTE DEPT.	18.1	24'	23'	790	NESBITT ARCH SILL LINE – STYLE GN12	UNFINISHED	INTERNAL SLEEVE AT MULLIONS	
R/2	VAULT OFFICE	11.8	15-1/2'	15'	790	NESBITT ARCH SILL LINE – STYLE GN12	FINISHED	INTERNAL SLEEVE AT MULLIONS	
R/3	LOAN DEPT.	24.5	32'	31'	790	NESBITT ARCH SILL LINE – STYLE GN12	FINISHED	INTERNAL SLEEVE AT MULLIONS	
R/4	OFFICE	6.5	15'	7'	975	NESBITT ARCH SILL LINE – STYLE FBN-14	UNFINISHED	NONE	ENCLOSURE FULL LENGTH OF WALL
R/5	TELLERS	27.6	36'	35'	790	NESBITT ARCH SILL LINE – STYLE GN12	FINISHED	INTERNAL SLEEVE AT MULLIONS	
R/6	TOILET	2.1	3'	3'	780	NESBITT ARCH SILL LINE – STYLE G-10A	UNFINISHED	NONE	

Fig. 7-9.　Radiation schedule.

radiator and second, to aid the workmen on the job in "roughing-in" for the piece of equipment.

Active Length   The active length is the actual length of the radiation tubing inside the housing, and it is given for the same reasons described under Enclosure Length. Therefore, in radiator No. 1—which has an enclosure length of 24 feet and an active length of 23 feet—there are 6 inches of space (1 foot total) on each end of the radiation tubing for pipe connections.

Rating—Btu Lineal Foot   One reason this column is provided is that in case another brand of radiator is substituted, the heat output per lineal foot will be kept the same as indicated. Since radiator No. 1 (R/1) has an active length of 23 feet at 790 Btu/ft, the total head output is 23 × 790 = 18,170 Btu.

Prototype   This column gives the manufacturer's description of the unit used in the engineer's design, but another manufacturer's unit will usually be allowed, provided it is equal in capabilities to the one originally specified.

Type of Enclosure Back   A column of this type will appear only on certain jobs. The reason is that the units specified with an unfinished back will be placed against a solid wall where the back cannot be seen, while the finished backs will be installed at the bottom of a floor to ceiling glass wall where the backs of the unit can be seen from the outside.

Type of Sleeve   The type-of-sleeve column indicates the blank (nonactivated) connections between the radiator enclosures. Such sleeves are either inside or outside telescoping, as needed, and are adjustable up to 7 inches at each sleeve.

Remarks   The remaining column is used for added remarks necessary to clearly indicate what the engineer requires. In this schedule, radiator No. 4 (R/4) shall have an enclosure the full length of the wall on which it is to be installed. In some cases a "blank" extension may be necessary to acquire the desired length.

Cabinet Unit Heater Schedule

The cabinet unit heater schedule shown in Fig. 7-10 was used on HVAC working drawings to describe the nine different types of unit heaters installed in a bank building. Each of the unit heaters described in this schedule contained a hot-water coil and a fan to circulate the air over the coil. The hot-water coil received its heated water (180°F) from a centrally located hot-water boiler similar to the one described in the schedule in Fig. 7-6.

No.	SERVING	CFM	CAPACITY MBH	GPM	FAN MOTOR WATTS	FAN MOTOR V/Ø/H₂	TYPE	PROTOTYPE	REMARKS
				CABINET UNIT HEATER SCHEDULE					
UH 1	BASEMENT REAR ENTRANCE	200	15.0	1.0	110	120/1/60	RECESSED HORIZONTAL CLG. MTD.	TRANE MODEL E46 FORCE FLOW	
UH 2	REAR STAIRWELL	300	20.0	1.0	160	120/1/60	VERTICAL CABINET	TRANE MODEL B42 FORCE FLOW	BUILT-IN THERMOSTAT
UH 3	BASEMENT STORAGE	200	15.0	1.0	110	120/1/60	RECESS HOR CLG. MTD W/O.A.CONN	TRANE MODEL E46 FORCE FLOW	PROVIDE BRICK VENT FOR O.A.
UH 4	BASEMENT VAULT	200	15.0	1.0	110	120/1/60	HORIZONTAL EXPOSED CAB CLG. MTD.	TRANE MODEL D16 FORCE FLOW	
UH 5	BASEMENT ELEVATOR LOBBY	200	15.0	1.0	110	120/1/60	VERTICAL CABINET	TRANE MODEL B42 FORCE FLOW	BUILT-IN THERMOSTAT
UH 6	MAIN STAIRWELL	400	30.0	1.5	150	120/1/60	VERTICAL CABINET	TRANE MODEL B42 FORCE FLOW	BUILT-IN THERMOSTAT
UH 7	MAIN VAULT	200	15.0	1.0	110	120/1/60	HOR. EXP. CABINET CLG. MTD	TRANE MODEL D16 FORCE FLOW	
UH 8	ENTRANCE	400	30.0	1.5	150	120/1/60	RECESSED HORIZONTAL CLG. MTD.	TRANE MODEL E46 FORCE FLOW	
UH 9	BOOK VAULT	200	15.0	1.0	110	120/1/60	HOR. EXP CABINET CLG. MTD	TRANE MODEL D16 FORCE FLOW	

Fig. 7-10.   Cabinet unit heater schedule.

Summary

In general, a schedule is a systematic method of presenting notes or lists of materials, components, equipment, and the like on a drawing in tabular form. When properly organized and thoroughly understood, they are powerful time-saving methods for the workmen on the job.

Questions

1. In the material schedule in Fig. 7-2, how many cubic yards of fill gravel is needed for the garage and terraces?
2. In the lighting fixture schedule in Fig. 7-3, what information does the "mounting" column give?
3. If schedules were not included on the drawings, where else would the information have to appear under normal circumstances?
4. In the grille schedule in Fig. 7–7, what does the term "CFM" mean?
5. In the radiation schedule in Fig. 7-9, what does the R/1 mean?
6. How many different types of cabinet unit heaters are listed in the schedule in Fig. 7-10?

Answers to Questions

1. 42 yd^3
2. The way the fixtures are mounted; that is, recessed in the ceiling, on the wall, etc.
3. In the written specification.
4. Cubic feet per minute.
5. Identifier Nesbitt Arch Sill Line—Style GN12.
6. 9

8 Site Plans

Objectives

To describe building site plans and methods of interpreting them and to show how the engineer's scale is used to measure dimensions on site plans.

A site plan is a plan view (as if viewed from an airplane) that shows the property boundaries and the building(s) drawn to scale and in its (their) proper location on the lot. Such plans will also include (where applicable) sidewalks, drives, streets, and similar details. Utilities such as water lines, sanitary sewer lines, telephone lines, and electrical power lines must also appear on site plans—sometimes on the original site plan furnished by the architect and other times on a separate site plan prepared by the engineering firm.

In actual practice, the initial site plan will be prepared by a certified land surveyor from information obtained from a deed description of the property. This property survey, however, will show only the property lines and their lengths as if the property were perfectly flat. If additional information is necessary, a complete field or topographic survey will have to be made. This topographic survey, in addition to showing the property lines, will also show the physical characteristics of the land by using contour lines, notes, and symbols. The physical characteristics may include:

- all property lines
- pertinent landmarks
- the direction of the land slope
- whether the land is flat, hilly, wooded, swampy, high, or low, and other features of its physical nature.

Figure 8-1 shows a typical site plan of a project as it might be drawn by an architect. This plan is complete in that it shows the property lines, the existing contour lines, the new contour lines (after grading), the location of the buildings on the property, new and existing roadways, all utility lines, etc. Descriptive notes appearing on the plan also

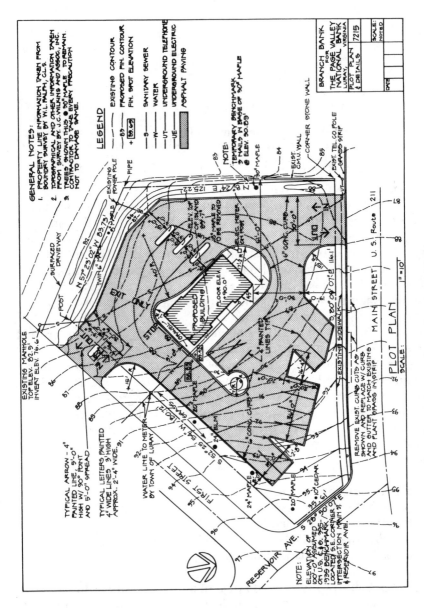

Fig. 8-1. Typical site plan of a project as drawn by an architect.

list the names of adjacent property owners, the land surveyor and the date of the survey, the direction of north, and other pertinent data, such as bench marks, monuments, and the like. A legend or symbol list is also provided so that anyone familiar with site plans can readily read the information provided. This is a high-quality site plan (called "Plot Plan" on the drawing), and the reader should study this plan until he is thoroughly familiar with all details described on it.

Using the Engineer's Scale

Earlier in this book, we found that most buildings are drawn to scale with a scale known as the architect's scale; its use was fully explained earlier. Site plans are drawn to scale also, but in most instances, a scale known as the "engineer's scale" is used rather than the architect's scale, except that an inch is divided into 10, 20, 30, 40, 50, or 60 equal parts (Fig. 8-2). The engineer's scale marked 10 is read 1" = 10' or some other integral power of 10; for example, 1" = 100'; 1" = 1000'; etc.

Usually, for small buildings on small lots, a scale of 1" = 10' or 1" = 20' is used. This means that 1 inch (actual measurement) on the drawing is equal to 10 or 20 feet—whichever the case may be—on the actual lot. Since the engineer's scale is the chief means of making scaled site plans, its use should be thoroughly understood.

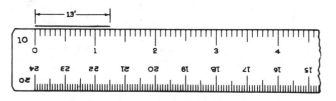

Fig. 8-2.   Typical engineer's scale.

The engineer's scale is used by placing it on the drawing with the working edge away from the user, as shown in Fig. 8-3. The scale is then aligned in the direction of the required measurement, such as along a property line. Then, looking down over the scale, the user reads the dimension, in the case of an existing drawing, or marks off the required dimension, in the case of a line that is to be drawn. The purpose of this scale, therefore, is to transfer the relative dimensions of an object to the drawing, or vice versa.

Although the drawing itself may appear reduced in scale, depending on the size of the object and the size of the sheet to be used, the actual true-length dimensions must be shown on the drawings at all times. When reading or drawing plans to scale, they should be thought of as "full size," and not in the reduced size that it happens to be on the

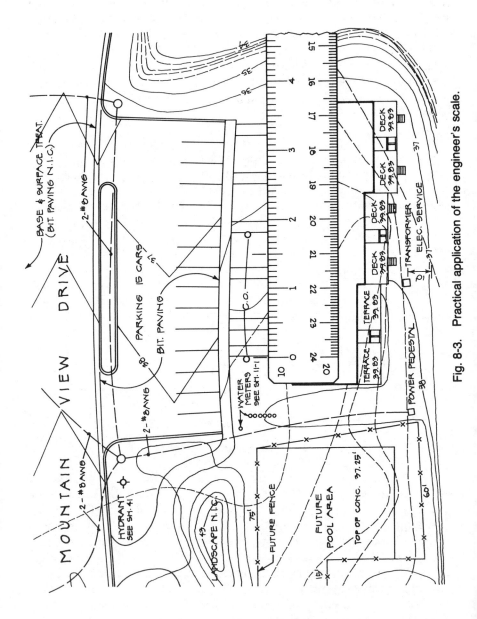

Fig. 8-3. Practical application of the engineer's scale.

drawing. For example, if you use a scale of 1" = 10' and the actual measurement on the drawing happens to be 3½", you do not say that the particular line is three and one-half inches; rather, read the measurement as thirty-five feet (3½" × 10 = 35).

The practice problems in Fig. 8-4 will acquaint you with the use of the most commonly used graduations on the engineer's scale. For each problem, use the scale indicated just below the line. Determine the length of each line and write it just above the line. When all the lengths have been determined, compare your answers with the ones given at the end of this chapter.

If it is desirable to draw a given line to a given scale, first mark off the distance with the appropriate scale; this is indicated by two light dots on the drawing. Then use a straightedge to draw the line between the dots.

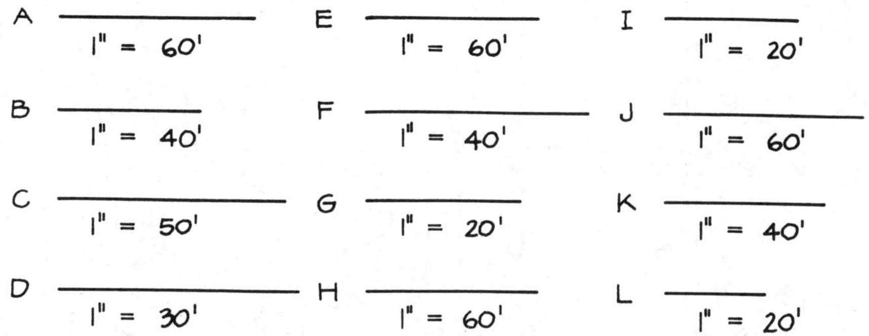

A ——————————
  1" = 60'

E ——————————
  1" = 60'

I ——————
  1" = 20'

B ——————
  1" = 40'

F ————————————
  1" = 40'

J ————————————
  1" = 60'

C ——————————
  1" = 50'

G ——————
  1" = 20'

K ——————————
  1" = 40'

D ————————————
  1" = 30'

H ——————————
  1" = 60'

L ————
  1" = 20'

Fig. 8-4. Practice problems to acquaint you with the use of the engineer's scale.

Refer to the site plan in Fig. 8-5 and note that the scale for this drawing is 1" = 40'. This means that every inch (actual measurement on the drawing) is equal to 40 feet on the building site. Every drawing should have the scale to which it is drawn plainly marked upon it either as shown in Fig. 8-5 or as a part of the title block.

With a site plan drawn to any given scale (1" = 20', 1" = 100', etc.) water and sewer lines, roadways, sidewalks, fire hydrants, and other site work may be described on the site plan to show the workmen exactly where these items are to be placed in relation to the building and the property lines. The site plan also shows the location of the building on the site in relation to the property lines, so that footings and foundations lines can be located as designed.

Other Scales

Mechanical engineer's scale: The mechanical engineer's scale in Fig. 8-6 is typical of those in current use. On these scales the major end unit

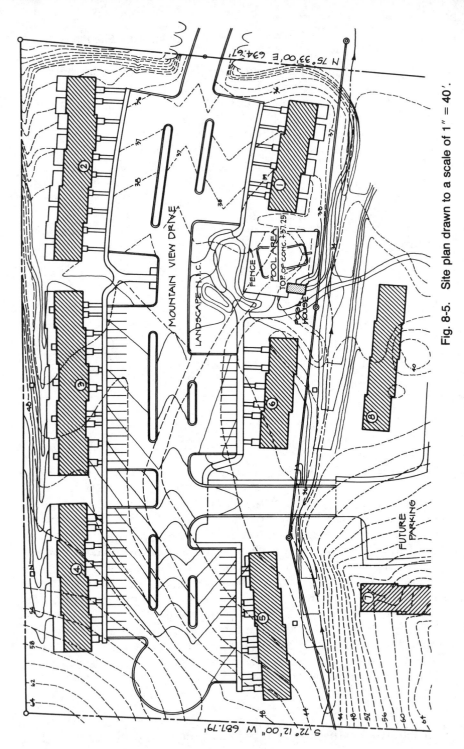

Fig. 8-5. Site plan drawn to a scale of 1″ = 40′.

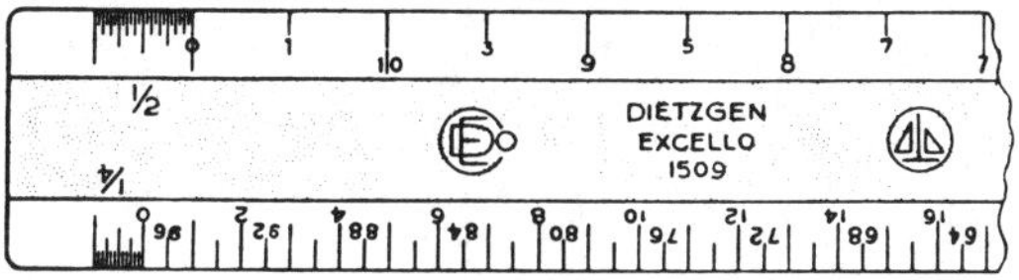

Fig. 8-6. Typical mechanical engineer's scale.

represents 1 inch and the subdivisions represent the commonly used fractions of an inch. These fractions are almost always in multiples of ½; that is, ½, ¼, ⅛, 1/16, etc. Such fractions as ⅓, ⅕, ⅙, etc. or other odd numbers are seldom, if ever, used.

Decimal scale: The decimal scale in Fig. 8-7 is designed with decimal dimensioning rather than fractional. In such cases, two-place decimals are normally used. They are made in even numbers, so that when halved, a two-place decimal results.

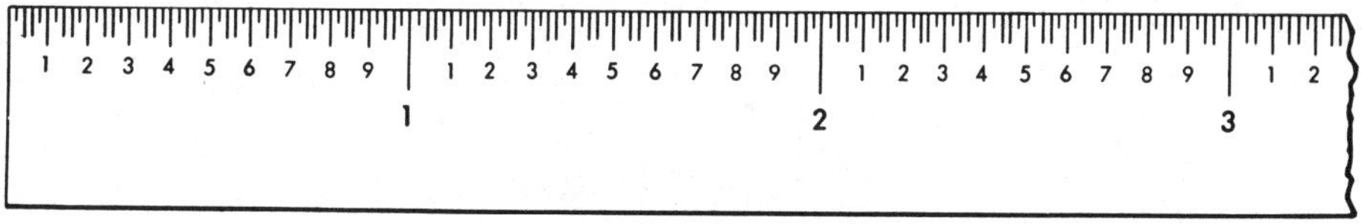

Fig. 8-7. Decimal scale.

The civil engineer's scale, discussed earlier in this chapter, is actually a form of decimal in that graduations on this scale represent decimals units. The engineer's scale is divided into 10, 20, 30, 40, 50, etc. parts of an inch—again in even numbers. If the edge is marked 20, this means that one inch has been subdivided into 10 equal parts. The scale most often used in engineering drawings is the one where an inch is divided into 50 equal parts. Each part, therefore, is 1/50 of an inch, or a decimal equivalent of .02 inch.

There are also scales like the ones previously discussed which use the metric system of measuring. These scales will more than likely eventually replace the inch, fractions, and decimals of an inch.

All scales are either open divided or fully divided. Open divided scales are those on which the main units are numbered along the whole length of the edge with an extra unit fully subdivided in the opposite direction from the zero point. The subdivided unit shows the fractional graduations of the main unit. Open divided scales often have two complete measuring systems on one face. For example, ¼″ and ⅛″ scales may appear on one face of the architect's scale as shown in Fig. 8-8.

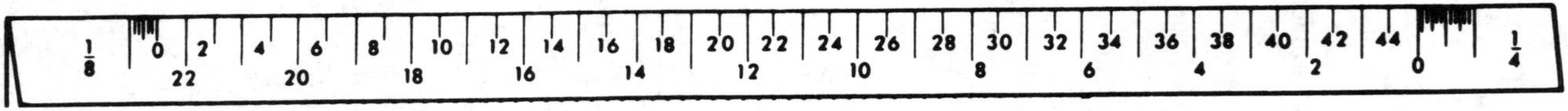

Fig. 8-8. Architect's scale with ¼" and ⅛" scales appearing on one face.

Fully divided scales have all the subdivisions along the entire length of the ruler so that several values from the same origin can be read without having to reset the scale. This is accomplished by double numbering, either to permit both right-to-left and left-to-right readings, or to provide two different scales on one face. See Fig. 8-9.

Protractors

Protractors are another type of scale used in blueprint reading, although not as often used in building construction drawings as the scale previously mentioned.

The protractor is a device for laying off and measuring angles on drawings. The best form for use on building construction drawings consists of a full circle graduated in degrees or fractions of a degree.

A half circle protractor is shown in Fig. 8-10. To lay off an angle, the center "0" of the protractor is placed at the vertex of the angle with the edge of the bar coinciding with the "1" line to which the angle is referred. A light mark is then made on the drawing at the proper graduation of the protractor arc. After the protractor is removed, a line is drawn joining this mark with the vortex. Angles can be measured in a similar manner.

Fig. 8-9.   Fully divided scale utilizing double numbering.

Fig. 8-10.   Half circle protractor.

Summary

A site plan of a building construction site is a plan view of the property showing all boundaries, and the location of the building on the property including sidewalks, driveways, and similar details. Utilities are normally shown also. These may include water lines, sanitary sewer, electrical, telephone, and so forth.

Site plans are drawn to scale, normally using the civil engineer's scale as a basis for dimensioning.

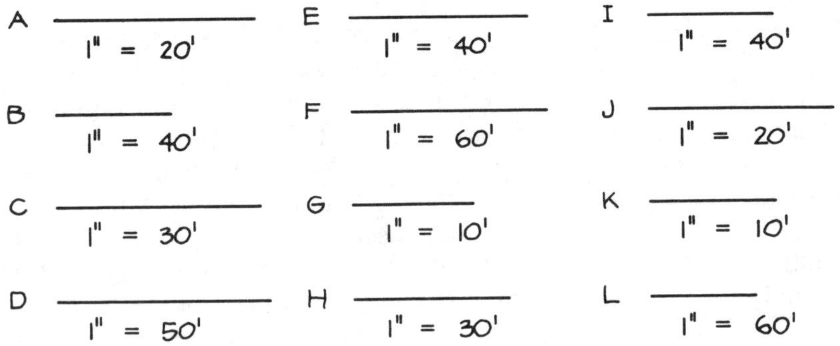

A ———————————   E ———————————   I ———————
    1" = 20'   1" = 40'   1" = 40'

B ——————   F ———————————   J ———————————
    1" = 40'   1" = 60'   1" = 20'

C ———————————   G ————————   K ——————————
    1" = 30'   1" = 10'   1" = 10'

D ———————————   H ————————   L ————————
    1" = 50'   1" = 30'   1" = 60'

Fig. 8-11.   Practice lines for using the engineer's scale.

Questions

1. The lines in Fig. 8-11 are of varying lengths and also drawn to various scales. Using your civil engineer's scale measure each line, using the proper scale as noted, and fill in the dimensions in the spaces provided.
2. What is normally placed on a set of construction documents so that anyone can read the symbols used on the drawing?
3. Name four different utility systems that are normally shown on building site plans.
4. Who normally prepares site plans initially?
5. Explain the difference between open divided scales and fully divided scales.

Answers to Questions

1.   A. 20 ft.
     B. 26 ft.
     C. 30 ft.
     D. 56 ft.

E. 35 ft.
F. 58 ft.
G. 6 ft.
H. 23 ft.
I. 25 ft.
J. 19 ft.
K. 6 ft.
L. 30 ft.

2. A symbol list or legend.
3. Electrical, plumbing, telephone, and cable television.
4. Land surveyor.
5. Fully divided scales are the subdivisions along the entire length of the ruler so that several values from the same origin can be read without having to reset the scale. Open divided scales are those on which the main units are numbered along the whole length of the edge with an extra unit fully subdivided in the opposite direction from the zero point.

9 Construction Specifications

Objectives

To become familiar with written specifications used with construction documents.

The construction specifications for a building construction project are the written descriptions of work and duties required of the architect, engineer, or owner. Together with the working drawings, these specifications form the basis of the contract requirements.

Sections of the Specifications

Divisions 1 through 16 of the written specifications cover requirements of a specific part of the construction work on the project. Included in these divisions are the type and grade of materials to be used, equipment to be furnished, and the manner in which equipment and components are to be installed. Each division will indicate the extent of the work covered and should be so written as to leave absolutely no doubt in anyone's mind of whether a certain part of the work—to be performed by a certain specialty contractor—is included in one section of the specifications or another.

The following is an outline of the various sections normally included in a complete set of construction specifications.

DIVISION 1—GENERAL REQUIREMENTS. This division covers a summary of the work, alternatives, project meetings, submittals, quality control, temporary factilities and controls, products, and the project closeout. Every responsible person involved with the project should become familiar with this division.

DIVISION 2—SITE WORK. Usually this division covers such items as foundation drains, porous fill, underground utilities, and other items related to areas outside of the building.

DIVISION 3—CONCRETE. Work covered under this division includes concrete formwork, expansion and contraction joints, concrete reinforcement, cast-in-place concrete, specially finished concrete, specially placed concrete, precaste concrete, and cementitious decks.

DIVISION 4—MASONRY. Covers mortar, masonry accessories, unit masonry, stone, masonry restoration and cleaning, as well as refactories.

DIVISION 5—METALS. Structural metal framing, metal joists, metal decking, lightgage framing, metal fabrications, ornamental metal, and expansion control normally fall under this division of the construction specifications.

DIVISION 6—CARPENTRY. Most items pertaining to wood fall under this division: rough carpentry, heavy timber construction, trestles, prefabricated structural wood, finish carpentry, wood treatment, architectural woodwork, and the like. Plastic fabrications are also included if used on the project for which the specifications are written.

DIVISION 7—THERMAL AND MOISTURE PROTECTION. The description of items in this division cover such items as waterproofing, dampproofing, building insulation, shingles and roofing tiles, preformed roofing and siding, membrance roofing, sheet metal work, wall flashing, roof accessories, and sealants.

DIVISION 8—DOORS AND WINDOWS. Metal doors and frames, wood and plastic doors, special doors, entrances and storefronts, metal windows, wood and plastic windows, special windows, hardware and specialties, glazing, and window wall/curtin walls are all covered in this division of the written specifications.

DIVISION 9—FINISHES. This division gives the types, quality, and workmanship of lath and plaster, gypsum wallboard, tile, terrazzo, acoustical treatment, ceiling suspension systems, wood flooring, resilient flooring, carpeting, special flooring, floor treatment, special coatings, painting, and wall covering.

DIVISION 10—SPECIALTIES. Specialty items such as chalkboards and tackboards, compartments and cubicles, louvers and vents that are not connected with the heating ventilating and air-conditioning contract, wall and corner guards, access flooring, specialty modules, pest control, fireplaces, flagpoles, identifying devices, pedestrian control devices, lockers, protective covers, postal specialties, partitions, scales, storage shelving, wardrobe specialties, and the like are covered in this division.

DIVISION 11—EQUIPMENT. The equipment included in this division are vacuum cleaning systems, and bank and vault, commercial, checkroom, darkroom, educational, food service, vending, athletic, industrial, laboratory, and laundry equipment, just to name a few.

DIVISION 12—FURNISHING. Items covered in this division include artwork, cabinets and storage, window treatment, fabrics, furniture, rugs and mats, seating, and other similar furnishing accessories.

DIVISION 13—SPECIAL CONSTRUCTION. Such items as air-supported structures, incinerators and other special items will fall under this division of the specifications.

DIVISION 14—CONVEYING SYSTEMS. As the name implies, this division covers conveying apparatus such as dumb-waiters, elevators, hoists and cranes, lifts, material-handling systems, turntables, moving stairs and walks, pneumatic tube systems, and powered scaffolding.

DIVISION 15—MECHANICAL. In general, this division of the construction specifications covers all the work of the heating, ventilating, air-conditioning, and plumbing contractors.

DIVISION 16—ELECTRICAL. This division covers all of the interior and exterior electrical work for the project.

Interpreting Construction Specifications

In general, construction specifications give the grade of materials to be used on the project and the manner in which they are to be installed. Most specification writers use an abbreviated language; although it is relatively difficult to understand at first, experience makes possible a proper interpretation with little difficulty. However, before any of the workmen begin a project, or portion thereof, they should make certain that everything is clear. If it is not, the architectural or engineering firm should be contacted to clarify the problem prior to beginning the work, not after the job is completed.

Most architects and consulting engineers will specify the type and design of certain items for the project by listing a given manufacturer's name and catalog number (if available) of the item. However, when items are specified this way, usually an item that is considered *an approved equal* will be accepted.

Some architects, engineers, and owners will be very reasonable when it comes to approving other than specified makes of components and equipment; others may require such close compliance that no other manufacturer's item will be accepted as an "equal." This is especially true when the architect/engineer allows a particular manufacturer to write a portion of the specifications for the firm. In this case, the manufacturer is certain to write an iron-clad specification which only his (the manufacturer's) equipment will meet.

Common Conflicts

Those involved with the building construction industry in any capacity should always be on the alert for conflicts between working drawings and the written specifications. Such conflicts occur particularly when:

1. Architects or engineers use standard or prototype specifications and attempt to apply them to specific working drawings without any modification.

2. Previously prepared standard drawings are to be changed or amended by reference in the specifications only; the drawings themselves are not changed.

3. Items are duplicated in both the drawings and specifications and then an item is amended in one and overlooked on the other contract document.

In such instances, it is the responsibility of the person in charge of the project to ascertain which takes precedent over the other; that is, the drawings or the specifications. When such a condition exists, the matter must be cleared up immediately, preferably before the work is installed, in order to avoid added cost to either the owner, architect/engineer, or the contractor.

Sample Specification

Construction specifications can vary in size from only a page or two to volumes containing hundreds of pages. To give the reader an overall view of a set of specifications, the following is a condensed version of a set of specifications used on a actual project. The exercises at the end of this chapter are designed to help the reader learn to use written specifications with ease.

DIVISION 1   GENERAL REQUIREMENTS

SECTION 1A   SPECIAL CONDITIONS

1. GENERAL:

A. The work of this contract includes the furnishing of all labor, materials, plant, equipment, tools and services for the construction of the Western Virginia Bicentennial Information Center, Albermarle County, Virginia, complete, and ready to occupy in accordance with the terms of the Agreement, General Conditions, Supplementary Conditions and Division 1 General Requirements.

B.   The work of this contract includes all items of construction, all plumbing, mechanical, electrical installation and other equipment as shown or specified; except as otherwise provided hereafter.

DIVISION 2   SITE WORK

SECTION 2B   EARTHWORK

The General Conditions, any Supplementary General Conditions, and Division 1, General Requirements, are hereby made a part of this section as fully as if repeated herein.

1.   GENERAL: Furnish all materials and equipment, and perform all labor necessary to do all excavating, filling and grading within the contract limits as called for by the drawings and specifications.

2.   WORK INCLUDED IN THIS SECTION:

   A.   Excavating

   B.   Filling and backfilling

   C.   Grading

   D.   Compaction

   E.   Testing

3.   WORK NOT INCLUDED IN THIS SECTION:

   A.   Roads, curbs, gutters and walks

   B.   Excavation and backfilling for mechanical and electric work

   C.   Seeding, sodding and landscaping

4.   EXISTING SITE CONDITIONS:

   A.   The Contractor shall visit the site and familiarize himself with the existing conditions and limitations of the work.

5.   EXCAVATION:

   A.   After all stripping has been completed, excavation of every description, regardless of material encountered, within the grading limits of the project shall be performed to the lines and grades indicated. Satisfactory excavated material shall be transported to and placed in fill areas within the limits of the work. When directed, unsatisfactory material encountered within the limits of the work shall be excavated below the grade shown and replaced with satisfactory material as directed. Such material excavated and the selected material ordered as replacement shall be included in excavation. Unsatisfactory excavated material shall be disposed of within the areas designated on the plans for excess fill material at the Contractor's expense and responsibility. During construction, excavation and filling shall be performed in a manner and sequence that will provide drainage at all times. Material required for fills in excess of that produced by excavation or cutting within the grading limits shall be excavated from approved borrow areas selected by the Contractor, as specified herein.

DIVISION 3   CONCRETE

SECTION 3A   CAST-IN-PLACE CONCRETE

The General Conditions, any Supplementary General Conditions, and Division 1, General Requirements, are hereby made a part of this section as fully as if repeated herein.

1. GENERAL:

A. Unless otherwise noted or specified all work shall conform to the applicable portion of the latest edition of the following Standard Specifications, which shall be available to the Contractor at the job site at all times during concreting:

1. Recommended Practice for Concrete Formwork (ACI 347-68)
2. Recommended Practice for Measuring, Mixing, and Placing Concrete (ACI 614-59)
3. Recommended Practice for Cold Weather Concreting (ACI 306-66)
4. Recommended Practice for Hot Weather Concreting (ACI 605-59)
5. Recommended Practice for Placing Reinforcing Bars (CRSI 63)
6. Recommended Practice for Placing Bar Supports, Specification and Nomenclature 1968 (CRSI-68)
7. Building Code Requirements for Reinforced Concrete (ACI 318-71)

8.   Latest Edition of all ASTM Specifications mentioned herein

2.   WORK INCLUDED:

A.   The work to be performed under this Section of the Specifications comprises the furnishing of all labor and materials and the completion of all work of this Section as shown on the Drawings and/or herein specified.

B.   In general the work included under this Section consists of, but is not limited to, the following:

1.   Formwork
2.   Inserts and embedded steel plates and anchors
3.   Proportioning concrete
4.   Mixing concrete
5.   Tests
6.   Conditioning of subgrades
7.   Placing reinforcing
8.   Placing concrete
9.   Finishing concrete
10.   Curing concrete
11.   Joints
12.   Removal of forms
13.   Patching
14.   Hardeners
15.   Vapor barrier

3.   WORK NOT INCLUDED: In general the following related work is included in other Sections of the Specifications:

A.   Reinforcing steel and mesh (except placing)

B.   Precast architectural concrete

C.   Shop drawings

D.   Slab subgrades

E.   Perimeter insulation

F.   Waterproofing membranes

4.   WORK INSTALLED BUT PROVIDED BY OTHERS: The following embedded items will be provided by others, but shall be installed under this Section:

    A. Electric conduit

    B. Pipes and pipe sleeves

    C. Anchor bolts

DIVISION 3   CONCRETE

SECTION 3B   REINFORCING STEEL

The General Conditions, any Supplementary General Conditions, and Division 1, General Requirements, are hereby made a part of this section as fully as if repeated herein.

1. GENERAL: Furnish all labor and materials in connection with the fabrication and delivery of all reinforcing steel and accessories for Cast-In-Place Concrete, complete, in accordance with the drawings and as herein specified.

2. WORK NOT INCLUDED:

    A. The following related work is included in other sections of the Specifications:

        1. Concreting
        2. Placing of reinforcing and mesh
        3. Formwork

3. SHOP DRAWINGS:

    A. Submit shop drawings for all fabricated items in this Section. Drawings shall give all dimensions necessary for fabrication and placing of the reinforcing steel and accessories without reference to the project drawings.

4. APPLICABLE STANDARD SPECIFICATIONS:

    A. Reinforcing steel shall be fabricated and detailed in accordance with the "American Concrete Institute Building Code Requirements for Reinforced Concrete" (ACI 318-63) and unless otherwise shown on drawings in accordance with "Manual of Standard Practice for Detailing Reinforced Concrete Structures" (ACI 315-65).

5. MATERIALS:

    A. Reinforcing Bars: Deformed billet steel bars, meeting ASTM A615, Grade 60. Bars shall be mill marked to identify stress grade.

Bars for stirrups and ties shall be intermediate grade, meeting ASTM A615, Grade 40. All bars to be deformed in accordance with ASTM A615-68.

B.   Welded steel wire fabric shall conform to ASTM A185-64.

C.   Metal Accessories: As required for rigid placement of reinforcing steel. Note special accessory requirements, if any, listed under "Cast-In-Place Concrete Work", Section 3A, Paragraph 13C. Bar supports placed on insulating material shall have spreader plates to prevent indentation of the insulation.

> 1.   Sheared length: ± inch
> 2.   Depth of trussed bars: + 0——−½″
> 3.   Stirrups, ties and spirals: ± ½″
> 4.   All other bends: ± 1″

DIVISION 4   MASONRY

SECTION 4A   MORTAR

The General Conditions, any Supplementary General Conditions, and Division 1, General Requirements, are hereby made a part of this section as fully as if repeated herein.

1.   GENERAL: Furnish all mortar materials for unit and stone masonry.

2.   MATERIALS:

A.   All materials shall be so delivered, stored and handled to prevent inclusion of foreign materials or damage by water or breakage. Materials not meeting specified requirements shall be removed from the premises immediately, and the Contractor shall furnish materials of the quality specified or shown on the drawings before proceeding with the work. If requested, the Contractor shall submit testing data as evidence of compliance with furnishing quality of materials specified. The report shall be prepared by a recognized and qualified testing laboratory.

B.   Portland Cement, natural Gray color, shall be Type I, as specified, in ASTM Specification C150, delivered in original containers, and shall bear maker's name and show batch number. One brand of cement shall be used throughout the work unless otherwise approved by the Architect.

C. Hydrated lime shall be masons' hydrated finishing lime and shall comply with Standard Specifications of ASTM C207 latest edition, Type S.

D. Masonry Cement, a ready-mixed mortar, shall conform to ASTM Specification C91, Type II. Mortar shall be as manufactured by the Riverton Lime Company or approved equal. Color: C-74. Mortar shall be non-staining. It is the intent of this specification that the sand and mortar used shall match that used at the Piedmont Virginia Community College.

E. Sand for mortar shall be of an acceptable color, clean, conforming to ASTM Specification C144, latest edition and graded so that 100% shall pass #4 sieve and not more than 40% shall pass a #50 sieve.

F. Non-Staining Cement shall conform to ASTM C91-70.

G. Admixtures, other than anti-freeze compounds, may be used in the mortar subject to approval of the Architect. The admixture must not adversely affect the mortar bond or compressive strengths of mortar designed without use of admixture. The admixture shall not contain calcium chloride, chloride salts, or any other chemical that will be deleterious to metals embedded in the mortar.

3. EXECUTION:

A. MIXES: Mortar shall be freshly prepared and uniformly mixed in the proportions by volume conforming to ASTM C270, Type M, N. Proprietary mortars, if used, shall be mixed in strict accordance with the manufacturer's specifications.

B. MIXING: Mortar shall be mixed by placing ½ of the water and sand in a mechanical batch mixer. Then the cement, lime, and remainder of the water and sand shall be added. After all materials are in the mixer they shall be mixed for at least 3 minutes and long enough to make a complete intimate mix of all the materials.

C. TEMPERING: Mortar shall be tempered with water as needed to maintain the required consistency. Tempering on mortar boards shall be done only by adding water in a basin formed with the mortar. The mortar shall then be reworked into the water. Any mortar which is not used within 1½ hours of its initial mixing shall not be used.

4. USES: See provisions of other sections of this Division for specific uses and mixes for portions of the work.

DIVISION 5   METALS

SECTION 5A   STRUCTURAL STEEL

The General Conditions, any Supplementary Conditions, and Division 1, General Requirements, are hereby made a part of this section as fully as if repeated herein.

1. GENERAL:

A. All structural steel shall be in accordance with American Institute of Steel Construction "Specification for the Design, Fabrication and Erection of Structural Steel Buildings", adopted February 12, 1969, and "Code of Standard Practice for Steel Buildings and Bridges", revised July 1, 1970.

B. Welding shall be in accordance with the latest edition of the "Code for Welding in Building Construction" of the American Welding Society. Welders are to be certified in accordance with the qualification of welders by the American Welding Society, for the particular type welds they are making on this project.

2. WORK INCLUDED: Furnish all labor and materials to complete structural steel work indicated on the drawings and specified herein. See AISC Code of Standard Practice for definition of structural steel.

DIVISION 6   WOOD AND PLASTICS

SECTION 6A   ROUGH AND FINISH CARPENTRY

The General Conditions, any Supplementary General Conditions, and Division 1, General Requirements, are hereby made a part of this section as fully as if repeated herein.

1. GENERAL:

A. Furnish all labor and materials in connection with
    1. Rough carpentry, framing and bucks
    2. Nailers, grounds and blocking
    3. Wood furring
    4. Rough hardware
    5. Finish carpentry
    6. Installation of finish hardware

B. WORK INSTALLED, BUT PROVIDED BY OTHERS: The following items will be provided by others, but shall be installed under this section:

1. Architectural woodwork
2. Wood doors
3. Finish hardware

C. WORK NOT INCLUDED: Furnishing and erection of Prefabricated Structural Wood Trusses. See Section 6C.

D. APPLICABLE STANDARD SPECIFICATIONS: Unless otherwise noted or specified, all work shall conform to the applicable portion of the latest edition of the following Standard Specifications:

1. Western Wood Products Association Grading Rules
2. Southern Pine Inspection Bureau, Standard Grading Rules
3. American Plywood Association

2. MATERIALS:

A. LUMBER:

1. Soft wood lumber shall conform to the grading rules of the manufacturer's association under whose rules the lumber is produced.
2. Each piece of structural or framing lumber shall be marked with grade, trademark and mill mark.
3. All lumber shall be air dried and well seasoned.
4. Light framing, joists and planks shall be 1500F, No. 1 or construction grade Yellow Pine, Douglas Fir or Western Hemlock, S4S finish.
5. Sleepers, grounds, furring and blocking shall be 1200F, No. 2 or Standard Grade Yellow Pine, Douglas Fir or Western Hemlock, S4S finish.

B. ROUGH HARDWARE: Provide all rough hardware items required to permanently secure all rough carpentry. Items shall include all types and sizes of nails, screws, nuts and bolts, washers, anchors and other items.

DIVISION 7   THERMAL AND MOISTURE PROTECTION

SECTION 7A   WATERPROOFING

The General Conditions, any Supplementary General Conditions, and Division 1, General Requirements, are hereby made a part of this section as fully as if repeated herein.

1. GENERAL:

A. Furnish all labor and materials in connection with waterproofing and dampproofing of walls and floors, complete, in accordance with the drawings and as herein specified.

B. Waterproofing shall be applied to the following surfaces and elsewhere as designated on the drawings:

1. On below grade masonry and/or concrete surrounding occupied spaces.
2. Between finish paving of Terrace and structural slabs where not supported on grade.

2. SHOP DRAWINGS: Submit samples of all materials proposed for use with certification of conformity with specifications.

3. WATERPROOFING:

A. MATERIALS:

1. All areas so designated shall be waterproofed with a minimum .060″ membrane of rubberized asphalt integrally bounded to polyethylene sheeting. Bituthene manufactured by W. R. Grace and Co., or equal.
2. The material shall conform to the following requirements:
Permenace—Perms   0.1 max. ASTM E96
Pliability: 180 degrees
        Bend over ¼″
        Mandrel at −50 degrees F.
        No cracks
        ASTM D146
Peel adhesion: 7 days dry
        + 7 days @ 120 degrees F.
        + 7 days water immersion
        (lb./in. width) 5.0 min.
        TT-S-00230 Modified
Crack bridging capability on application: ³/₁₆″

Cycling over crack at − 15 degrees F. (crack opened and closed from 0 to ¼″)
100 cycles min.
Puncture resistance: (lb.) 40 min. ASTM E154

3. Material shall be applied in strict accordance with the manufacturer's recommendations in addition to the general instructions noted herein.

DIVISION 7   THERMAL AND MOISTURE PROTECTION

SECTION 7B   INSULATION

The General Conditions, any Supplementary General Conditions, and Division 1, General Requirements, are hereby made a part of this section as fully as if repeated herein.

1. GENERAL: Furnish all labor and materials in connection with the installation of roof deck insulation and wall insulation, complete, in accordance with the drawings and as herein specified.

2. SAMPLES: Submit samples as listed hereafter:

One (1) 12″ × 12″ sample of each type in insulation proposed for use.

3. MATERIALS:

A. Roof Deck Insulation: Barrett Celo-Ther Insulation 1″ thick, with 0.36 Conductance (c) insulating value or approved equal.

B. Wall Insulation: Rigid thermal insulation shall be Styrofoam FR as manufactured by Dow Chemical Company or approved equal.

C. Insulation for sloped roof areas and elsewhere as indicated shall be Foil-faced Fiberglas Building Insulation or approved equal, with Thermal Resistance "R" value of R-19 or R-13 as noted on the details. Vapor barrier shall be not greater than Class 2 nor have flame spread rating of 75 or less.

4. INSTALLATION:

A. Roof deck insulation shall be applied over metal decks in a full mopping of Crystal Steep asphalt in strict accordance with Barrett specifications, and in addition must be in compliance with the Roofing Manufacturer's manual.

B. Wall insulation shall be installed horizontally within cavity space against inner wythe of back-up block as detailed. Installation shall be in strict accordance with manufacturer's published specifications. Insulation shall be secured in place by wall ties supplemented by mastic as required.

C. Batt insulation shall be installed as noted and detailed between rafters or bottom chord of trusses and in accordance with the manufacturer's recommendations.

D. Provide and install wire mesh or baling wire below insulation as recommended by the manufacturer to support insulation.

DIVISION 8   DOORS, WINDOWS AND GLASS

SECTION 8A   WOOD DOORS

The General Conditions, any Supplementary General Conditions, and Division 1, General Requirements, are hereby made a part of this section as fully as if repeated herein.

1. GENERAL:

A. Furnish all labor and materials in connection with wood doors as scheduled, complete, in accordance with the drawings and as herein specified.

B. Doors shall be of the type, size, and design shown. Top and bottom edges of the door shall be sealed with a clear water-resistant varnish or a clear water-resistant sealer prior to shipment. Doors shall be stored in fully covered, well ventilated areas and protected from extreme changes in temperature and humidity. Where shown, doors shall be prepared for the reception of glass and/or louvers.

2. FLUSH DOORS:

A. All flush veneered doors shall be Weldwood Novodor Doors with Novoply core as manufactured by U.S. Plywood or approved equal.

B. Veneers shall be as scheduled and hereafter specified.

1. Face veneer, unless otherwise noted, shall be Sliced White Oak, standard thickness, properly dried, laid with the grain at right angles to the grain of the cross bands. Both faces shall be smoothly sanded. Finish shall be toned Univar, factory applied.

C. All doors shall be prefit to net sizes and shall be factory machined for hardware as specified and scheduled.

3. SHOP DRAWINGS shall be submitted for approval in accordance with the Supplementary Conditions. Shop Drawings shall indicate the location of each door, elevation of each type of door, details of construction, marks to be used to identify the doors, location and extent of hardware blocking, and if factory-primed, materials and methods to be used. Shop Drawings shall include catalog cuts.

4. MARKING: Each door shall bear a stamp, brand, or other identifying mark indicating quality and construction of the door. The identifying mark or a separate certification shall include the name of the inspecting organization, identification of the standard on which the construction of the door is based, identity of the plant to which the stamp was issued, and a declaration of compliance by the plant.

5. INSTALLATION: Doors shall be installed only after completion of all other work which would raise the moisture content of the doors or damage the surface of the doors. Doors shall have a clearance at the bottom of ¼″ over thresholds and ½″ at other locations unless otherwise shown. The lock edge of doors shall be beveled at the rate of ⅛″ in 2″. Cuts made on the job shall be sealed immediately after cutting, using a clear water-resistant varnish or sealer.

6. GUARANTEE: All doors shall be guaranteed for the Life of Original Installation in accordance with the term sof the NWMA standard guarantee as supplemented to include finishes installed by the manufacturer.

DIVISION 9   FINISHES

SECTION 9A   LATH AND PLASTER

The General Conditions, any Supplementary General Conditions, and Division 1, General Requirements, are hereby made a part of this section as fully as if repeated herein.

1. GENERAL: Furnish all labor and materials in connection with the lathing and plastering, complete, in accordance with the drawings and as herein specified.

2. WORK NOT INCLUDED: The following related work is included in other sections of the specifications:

A. Metal stud and gypsum wallboard systems

B. Ceramic tile work

3. PROTECTION OF WORK: The Contractor shall protect his work and the work of others from damage due to his operations. The Contractor shall replace or repair, as required, all damaged work to the satisfaction of the Architect.

4. APPLICABLE STANDARD SPECIFICATIONS: Unless otherwise noted or specified, all work shall conform to the applicable portion of the latest edition of the following Standard Specifications:

ASA-A42.1   Specification for Gypsum Plastering
ASA-42.2 & 42.3   Portland Cement, Stucco and Plastering
ASA-42.4   Interior Lathing and Furring

DIVISION 10   SPECIALTIES

SECTION 10A   TOILET AND BATH ACCESSORIES

The General Conditions, any Supplementary General Conditions, and Division 1, General Requirements, are hereby made a part of this section as fully as if repeated herein.

1. GENERAL: Furnish all labor and materials in connection with the following toilet and bath accessories, complete, in accordance with the Drawings and as herein specified.

2. SHOP DRAWINGS: Submit shop drawings and installation instructions for all items specified herein.

3. TOILET AND BATH ACCESSORIES: Items specified and scheduled hereinafter are as listed and manufactured by Bobrick Washroom Equipment, Inc. Similar and equal products of other manufacturers may be substituted with the approval of the Architect.

A. Schedule of Accessories:

Item 1: *Recessed Paper Towel Dispenser and Waste Receptacle:* Model No. B-360P. Provide two (2) in each Men 107 and Women 108.

Item 2: *Surface-Mounted Paper Towel Dispenser:* Model No. B262. Provide one (1) in each Women 121 and Men 122.

Item 3:   *Surface-Mounted Stainless Steel Wall Urn:* Model B260. Provide one (1) at each urinal.

Item 4:   *Powdered Soap Dispenser:* Model No. B32. Provide one at each lavatory.

Item 5:   *Stainless Steel Framed Mirrors—Shelf Combination:* Model B290 1620. Provide One (1) over each lavatory.

Item 6:   *Partition Mounted Multi-Roll Toilet Tissue Dispenser:* Model B-386. Provide one in each toilet stall partition between toilets.

Item 7:   *Partition Mounted Napkin Disposal for Two Toilet Compartments:* Model B-354. Provide in each two toilet compartments for women.

Item 8:   *Stainless Steel Grab Bars:* Provide B-550 Series Wheelchair Toilet Compartment Grab Bars for the handicapped, B-593.

DIVISION 15   MECHANICAL

SECTION 15A   GENERAL PROVISIONS

Portions of the sections of the Documents designated by the letters "A", "B", & "C" and "DIVISION ONE—GENERAL REQUIREMENTS" apply to this Division. Consult Index to be certain that set of Documents and Specifications is complete. Report omissions or discrepancies to the Architect/Engineer.

1. SCOPE OF THE WORK: The scope of the work consists of the furnishing and installing of complete mechanical systems including miscellaneous systems. The Mechanical Contractor shall provide all supervision, labor, materials, equipment, machinery, and any and all other items necessary to complete the systems. The Mechanical Contractor shall note that all items of equipment are specified in the singular; however, the Contractor shall provide and install the number of items of equipment as indicated on the drawings and as required for complete systems.

A. It is the intention of the Specifications and Drawings to call for finished work, tested, and ready for operation.

B. Any apparatus, appliance, material or work not shown on drawings but mentioned in the specifications, or vice versa, or any incidental accessories necessary to make the work complete and perfect in all respects and ready for operation, even if not particularly specified, shall be furnished, delivered and installed by the Contractor without additional expense to the Owner.

C.   Minor details not usually shown or specified, but necessary for proper installation and operation, shall be included in the Contractor's estimate, the same as if herein specified or shown.

2.   SHOP DRAWINGS:

A.   The Mechanical Contractor shall submit five (5) copies of the shop drawings to the Architect for approval within thirty (30) days after the award of the general contract. If such a schedule cannot be met, the Mechanical Contractor may request in writing for an extension of time to the Architect. If the Mechanical Contractor does not submit shop drawings in the prescribed time, the Architect has the right to select the equipment.

3.   COOPERATION WITH OTHER TRADES:

A.   The Mechanical Contractor shall give full cooperation to other trades and shall furnish (in writing, with copies to Architect) any information necessary to permit the work of all trades to be installed satisfactorily and with least possible interference or delay.

4.   GUARANTEE:

A.   The Mechanical Contractor shall guarantee, by his acceptance of the contract, that all work installed will be free from defects in workmanship and materials. If during the period of one year, or as otherwise specified, from date of Certificate of Completion and acceptance of work, any such defects in workmanship, materials or performance appear, the Contractor shall, without cost to the Owner, remedy such defects within a reasonable time to be specified in notice from Architect. In default, the Owner may have such work done and charge cost to Contractor.

DIVISION 16   ELECTRICAL

SECTION 16A   GENERAL PROVISIONS

Portions of the sections of the Documents designated by the letters "A", "B", & "C" and "DIVISION ONE—GENERAL REQUIREMENTS" apply to this Division. Consult Index to be certain that set of Documents and Specifications is complete. Report omissions or discrepancies to the Architect.

1. SCOPE OF THE WORK:

A.   The scope of the work consists of the furnishing and installing of complete electrical systems—exterior and interior—including miscellaneous systems. The Electrical Contractor shall provide all supervision, labor, materials, equipment, machinery, and any and all other items necessary to complete the systems. The Electrical Contractor shall note that all items of equipment are specified in the singular; however, the Contractor shall provide and install the number of items of equipment as indicated on the drawings and as required for complete systems.

B.   It is the intention of the Specifications and Drawings to call for finished work, tested, and ready for operation.

C.   Any apparatus, appliance, material or work not shown on drawings but mentioned in the specifications, or vice versa, or any incidental accessories necessary to make the work complete and perfect in all respects and ready for operation, even if not particularly specified, shall be furnished, delivered and installed by the Contractor without additional expense to the Owner.

D.   Minor details not usually shown or specified, but necessary for proper installation and operation, shall be included in the Contractor's estimate, the same as if herein specified or shown.

E.   With submission of bid, the Electrical Contractor shall give written notice to the Architect of any materials or apparatus believed inadequate or unsuitable, in violation of laws, ordinances, rules; and any necessary items or work omitted. In the absence of such written notice, it is mutually agreed the Contractor has included the cost of all required items in his proposal, and that he will be responsible for the approved satisfactory functioning of the entire system without extra compensation.

2. ELECTRICAL DRAWINGS:

A.   The Electrical drawings are diagrammatic and indicate the general arrangement of fixtures, equipment and work included in the contract. Consult the Architectural drawings and details for exact location of fixtures and equipment; where same are not definitely located, obtain this information from the Architect.

B.   Contractor shall follow drawings in laying out work and check drawings of other trades to verify spaces in which work will be installed. Maintain maximum headroom and space conditions at all points. Where headroom or space conditions appear inadequate, the Architect shall be notified before proceeding with installation.

C.   If directed by the Architect, the Contractor shall, without extra charge, make reasonable modifications in the layout as needed to prevent conflict with work of other trades or for proper execution of the work.

3.   CODES, PERMITS AND FEES:

A.   Contractor shall give all necessary notices, including electric and telephone utilities, obtain all permits and pay all government taxes, fees and other costs, including utility connections or extensions, in connection with his work; file all necessary plans, prepare all documents and obtain all necessary approvals of all governmental departments having jurisdiction; obtain all required certificates of inspection for his work and deliver same to the Architect before request for acceptance and final payment for the work.

B.   Contractor shall include in the work, without extra cost to the Owner, any labor, materials, services, apparatus, drawings (in addition to contract drawings and documents) in order to comply with all applicable laws, ordinances, rules and regulations, whether or not shown on drawings and/or specified.

C.   Work and materials shall conform to the latest rules of the National Board of Fire Underwriters' Code, Regulations of the State Fire Marshal, and with applicable local codes and with all prevailing rules and regulations pertaining to adequate protection and/or guarding of any moving parts, or otherwise hazardous conditions. Nothing in these specifications shall be construed to permit work not conforming to the most stringent of applicable codes.

D.   The National Electric Code, the Local Electric Code, and the electrical requirements as established by the State and Local Fire Marshal, and rules and regulations of the power company serving the project, are hereby made part of this specification. Should any changes be necessary in the drawings or specifications to make the work comply with these requirements, the Electrical Contractor shall notify the Architect.

4. SHOP DRAWINGS:

A. The Electrical Contractor shall submit five (5) copies of the shop drawings to the Architect for approval within thirty (30) days after the award of the general contract. If such a schedule cannot be met, the Electrical Contractor may request in writing for an extension of time to the Architect. If the Electrical Contractor does not submit shop drawings in the prescribed time, the Architect has the right to select the equipment.

B. Shop drawings shall be submitted on all major pieces of electrical equipment, including service entrance equipment, lighting fixtures, panel boards, switches, wiring devices and plates and equipment for miscellaneous systems. Each item of equipment proposed, shall be a standard catalog product of an established manufacturer. The shop drawing shall give complete information on the proposed equipment. Each item of the shop drawings shall be properly labeled, indicating *the intended service of the material*, the job name and Electrical Contractor's name.

C. The shop drawings shall be neatly bound in five (5) sets and submitted to the Architect with a letter of transmittal. The letter of transmittal shall list each item submitted along with the manufacturer's name.

D. Approval rendered on shop drawings shall not be considered as a guarantee of measurements or building conditions. Where drawings are approved, said approval does not mean that drawings have been checked in detail; said approval does not in any way relieve the Contractor from his responsibility, or necessity of furnishing material or performing work as required by the contract drawings and specifications.

5. COOPERATION WITH OTHER TRADES:

A. The Electrical Contractor shall give full cooperation to other trades and shall furnish (in writing, with copies to Architect) any information necessary to permit the work of all trades to be installed satisfactorily and with least possible interference or delay.

B. Where the work of the Electrical Contractor will be installed in close proximity to work of other trades, or where there is evidence that the work of the Electrical Contractor will interfere with the work of other trades, he shall assist in working out space conditions to make a satisfactory adjustment. If so directed by the Architect, the

Electrical Contractor shall prepare composite working drawings and sections at a suitable scale clearly showing how his work is to be installed in relation to the work of other trades. If the Electrical Contractor installs his work before coordinating with other trades or so as to cause any interference with work of other trades, he shall make necessary changes in his work to correct the condition without extra charge.

C. The complexity of equipment and the variation among equipment manufacturers require complete coordination of all trades. The Contractor, who offers for consideration, substitutes of equal products of reliable manufacturers, has to be responsible for all changes that affect his installation and the installation and equipment of other trades. All systems and their associated controls must be completely installed, connected, and operating to the satisfaction of the Architect prior to final acceptance and contract payment.

6. TEMPORARY ELECTRICAL SERVICE:

A. The Electrical Contractor shall be responsible for all arrangements and costs for providing at the site, temporary electrical metering, main switches and distribution panels as required for construction purposes. The distribution panels shall be located at a central point designated by the Architect. The General Contractor shall indicate prior to installation whether three-phase or single-phase service is required.

7. ELECTRICAL CONNECTIONS:

A. The Electrical Contractor shall provide and install power wiring to all motors and electrical equipment complete and ready for operation including disconnect switches and fuses. Starters, relays and accessories shall be furnished by others unless otherwise noted, but shall be installed by the Electrical Contractor. This Contractor shall be responsible for checking the shop drawings of the equipment manufacturer to obtain the exact location of the electrical rough-in and connections for equipment installed.

B. The Mechanical Contractor will furnish and install all temperature control wiring and all interlock wiring unless otherwise noted.

C. It shall be the responsibility of the Electrical Contractor to check all motors for proper rotation.

8. AS-BUILT DRAWINGS:

A.   The Electrical Contractor shall maintain accurate records of all deviations in work as actually installed from work indicated on the drawings. On completion of the project, two (2) complete sets of marked-up prints shall be delivered to the Architect.

9. INSPECTION AND CERTIFICATES:

A.   On the completion of the entire installation, the approval of the Architect and Owner shall be secured, covering the installation throughout. The Contractor shall obtain and pay for Certificate of Approval from the public authorities having jurisdiction. A final inspection certificate shall be submitted to the Architect prior to final payment. Any and all cost incurred for fees shall be paid for by the Contractor.

10. TESTS:

A.   The right is reserved to inspect and test any portion of the equipment and/or materials during the progress of its erection. This Contractor shall test all wiring and connections for continuity and grounds, before connecting any fixtures or equipment.

B.   The Contractor shall test the entire system in the presence of the Architect or his engineer when the work is finally completed to insure that all portions are free from short circuits or grounds. All equipment necessary to conduct these tests shall be furnished at the Contractor's expense.

11. EQUIVALENTS:

A.   When material or equipment is mentioned by name, it shall form the basis of the Contract. When approved by the Architect in writing, other material and equipment may be used in place of those specified, but written application for such substitutions shall be made to the Architect as described in the Bidding Documents. The difference in cost of substitute material or equipment shall be given when making such request. Approval of substitute is, of course, contingent on same meeting specified requirements and being of such design and dimensions as to comply with space requirements.

12. GUARANTEE:

A.   The Electrical Contractor shall guarantee, by his acceptance of the contract, that all work installed will be free from defects in workmanship and materials. If during the period of one year, or as otherwise specified, from date of Certificate of Completion and acceptance of work, any such defects in workmanship, materials or performance appear, the Contractor shall, without cost to the Owner, remedy such defects within a reasonable time to be specified in notice from Architect. In default, the Owner may have such work done and charge cost to Contractor.

The reader should bear in mind that the sample specifications just presented are very condensed (due to the limited space available in this book) and that most written specifications for building construction will be much larger in content and will contain much greater detail of materials and installation methods, and the set will probably contain several hundred typewritten pages.

Summary

Specifications for a building or project are the written description of work and duties required of the architect, engineer, or owner. Together with the working drawings, these specifications form the basis of the contract requirements for the construction of the various systems for the building or project.

Division 1 through 16 of the written specifications cover requirements of a specific part of the construction work on the project. Included in these Divisions are the type and grade of materials to be used, equipment to be furnished, and the manner in which it is to be installed. Each Division will indicate the extent of the work covered and should be so written as to leave absolutely no doubt in anyone's mind whether a certain part of the work to be performed by a certain contractor is included in one section of the specifications or another.

Questions

1. Division 4 of the written specifications cover work under _____.

2. Under Division 4, all materials shall be so delivered, stored, and handled to prevent inclusion of _____ materials or damaged by _____ or _____.

3. All structural steel shall be in accordance with American Institute of _____.

4. Under waterproofing, all areas so designated shall be waterproofed with a minimum _____ inch membrane of rubberized asphalt.

5. _____ insulation shall be installed as noted and detailed

between rafters or bottom cord of trusses and in accordance with the manufacturer's recommendations.

6. Each door shall bear a _____, _____, or other identifying mark indicating quality and construction of the door.

7. Under Division 9, the following related work is included in other sections of the specifications:

   A. _____

   B. _____

8. The Mechanical Contractor shall submit _____ copies of the shop drawings to the Architect.

9. The Mechanical Contractor shall guarantee that all work installed will be free from defects and workmanship of materials for a period of _____ year(s).

10. It is the intention of the electrical specifications and drawings to call for _____ work, tested, and ready for operation.

Answers to Questions

1. masonry
2. foreign, water or breakage
3. Steel Construction
4. .060
5. batt
6. stamp, brand
7. A. metal stud and gypsum wallboard system
   B. ceramic tile work
8. 5
9. one
10. finished

10 Reproduction of Drawings

Objectives

To learn how drawings are reproduced for use by many workmen on a project.

A great amount of work is required to prepare even the simplest of drawings. Therefore, the original drawing of a building construction project is often very valuable and must be guarded against wear or becoming lost. One simple solution to this problem is to make copies of the original drawings for distribution to those who need them for reference.

Blueprinting

Blueprinting has been the accepted method of reproducing drawings from tracing paper for many years. Basically, the process of making a blueprint consists of placing a special paper that has been coated with chemicals sensitive to sunlight in a glass frame with its coated side up. The original drawing on tracing paper is laid over the coated paper, and both are held together by the glass frame.

The entire assembly is then subjected to sunlight for a few minutes, and then the coated paper is thoroughly washed in clean water and hung up to dry. If the exposure has been timed correctly, the coated surface of the paper is now a clear, dark blue color except where it was covered by the lines on the drawing; the latter are perfectly white.

An electric blueprinting machine is also used for making blueprints. In place of sunlight, the coated paper along with the original drawing passes around a glass cylinder containing electric lamps. The speed at which the paper travels may be adjusted to suit the quality of the drawings, as far as the intensity and size of the lines and the depth of the background are concerned.

Whiteprinting

Most drafting rooms now use the reproduction method known as whiteprinting. This method consists of exposing chemically treated paper and the original drawing to a high-intensity discharge or fluorescent lamp enclosed in a suitable housing. The ultraviolet light from either of these two light sources reduces the part of the sensitized surface that is unprotected by the lines of the original drawing into an invisible compound. After exposure, the print (or chemically treated paper) is then subjected to ammonia vapors, which develop the sensitized lines. The finished print is true to scale and ready for immediate use. Depending upon the type of paper used with this process, the printed lines may be red, brown, blue, or black on a white to light blue background.

Photocopying

Another convenient method of reproducing drawings (especially the smaller ones) is photocopying. This method actually takes a picture of the original drawing and then produces as many prints as desired—just like a Xerox or similar copying machine. This type of copier also has the advantage of making reductions or enlargements of the original drawing.

Microfilming

Over the past decade or so, microfilming has rapidly moved into the drafting industry for reproducing and preserving drawings. Because of the small size of the film, much space is saved, as well as shipping charges for the drawings.

The first step in microfilming a drawing is to convert the original drawing to a microfilm frame by means of a special camera mounted on a frame over a platform such as a table. The camera reduces a drawing so that it will fit on the microfilm; the reduction can be varied by changing the height of the camera above the drawing.

After the film is exposed, it is passed through a processor, which develops it, and it is then mounted on some type of card or frame.

Once the film has been mounted, the image can be blown up for viewing by means of a special viewer. Some viewers will even make full-size prints from the microfilm.

Since a great amount of reduction is obviously used in the microfilm process, drawings of the highest quality should be used, with every line sharp and the lettering of sufficient size.

Repair of Damaged Drawings

Sometimes original drawings become creased, stained, or worn to the point that they cannot be satisfactorily reproduced, revised, or microfilmed. Of course, the original drawing can be redrawn or traced, but often this involves quite a lot of work—running the expense high. There are photographic materials and techniques that can be used to restore old worn drawings to new usefulness. For example, the drawing in Fig. 10-1 is very worn—having creases and tears that have been repaired with transparent tape. Prints of this tracing will be hard to read. Now look at the drawing in Fig. 10-2. This is the same drawing as shown in Fig. 10-1 after it has been restored by photographic techniques.

One method of removing stains from drawings (or at least reduced) is by using filters over the lens of the photocopy camera. A filter is selected of a color as close to the color of the stain as possible; that is, a yellow stain would require a yellow filter; a red stain would require a red filter, etc. Original drawings that are in line can usually be copied so that the stain is eliminated since the filter will result in only a very faint image of the negative which will drop out when the print is made on high contrast paper. With continuous tone originals, success can be completely perfect only when the filter is closely matched to the color of the transparent stain. When in doubt as to which filter to select between a light, medium, or dark version of the particular color required, the denser color will do better.

Soiled tracings can usually be cleaned up satisfactorily by swabbing the surface with a wad of cotton that has been immersed in a solution consisting of equal quantities of water and alcohol. The cotton should be squeezed to remove surplus solution and then gently rubbed over the surface of the tracing.

Most construction drawings need to be revised or changed from time to time. Again, the photocopier can save valuable time. Rather than retrace a drawing that needs only a few changes, a photostat print can be made and the unwanted part cut out. After taping the remaining drawing to a new drawing form, the composite is photographed and printed on a special base film. The revisions can then be made on this film. Much time is saved by redrawing only those portions actually needing revision.

As mentioned previously, photocopiers can also be used to change the size of the original drawing. This is valuable when you want to reduce the size to save file space or cut printing and postage costs, or when you want to enlarge the drawing to open up detail clutter for greater legibility or to make revisions easier.

Other Miscellaneous Means of Copying Drawings

The methods previously described in this chapter are the ones most commonly used to reproduce drawings throughout the industry. Some

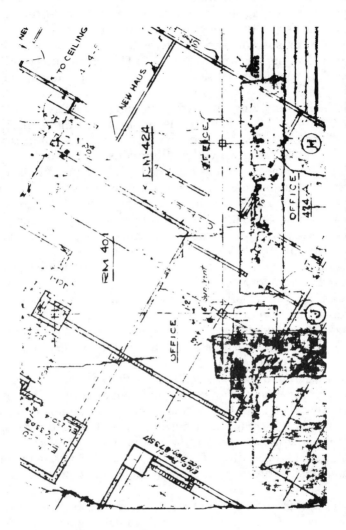

Fig. 10-1. A worn drawing that has creases and tears which have been repaired with transparent tape.

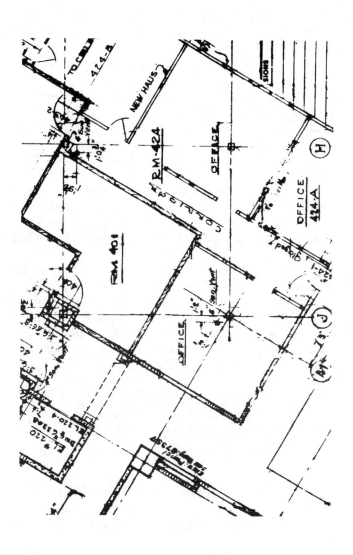

Fig. 10-2. The same drawing as shown in Fig. 10-1 after it has been restored by photographic techniques.

variations of these methods include thermography, ditto, mimeograph, diazo, xerograph, offset, etc.

Thermography is a process of reproducing drawings by means of infrared light. One copy of a drawing can be reproduced in about 6 seconds. The thermofax is a very easy machine to operate, but if more than one or two copies of a drawing are required, the process can be expensive.

The *ditto* process is a spirit duplicating method that uses a ditto master which can be sketched, drawn, or typed. The master is then attached to a duplicator drum and can make up to 100 prints directly from the master. The Ditto is seldom used in modern drafting rooms other than for duplicating temporary notes and sketches from 20 to 100 copies.

Where up to a thousand copies of drawings are required, the *mimeograph* machine is sometimes employed. A Stenafax machine can cut a stencil from an original drawing electronically, and the stencil is then mounted on the duplicating drum of the mimeograph machine for the run.

The *diazo* process of duplicating drawings is similar to blueline printing as discussed previously. The original tracing is laid on a piece of copy paper and then run through a machine to expose the copy paper to light. The parts of the paper beneath the lines and lettering of the drawing are not exposed to light and are developed into azo dye when passed over ammonia fumes. The color of the lines can be varied by changing the dye component—the three most normal colors being red, black, and blue.

The Multilith *offset* duplicating method of reproducing drawings is used when large-quantity production runs are required. The master is easily made and the resulting copies are very sharp and clear.

Summary

Due to the time involved to prepare even the simplest of drawings, some means of reproducing the drawings is necessary in order to distribute copies to the various building trades, inspection departments, and so forth.

Blueprinting has been the accepted method of reproducing drawings over the years. However, this process is rather time consuming compared to some of the more modern methods. Photographic reproduction and blueline (whiteprinting) are the two most used methods at the present time.

Questions

1. When large-quantity production runs of drawings are necessary, what method will give sharp and clear prints and would probably be the most practical method to use?
2. What copying method uses an azo dye to develop the lines and lettering on a drawing?
3. The Ditto process is sometimes used to reproduce up to 100 copies of a drawing. What similar method is used when up to 1000 copies of a drawing are desired?
4. What is the process that uses infrared light to reproduce one or two copies of a drawing?
5. What is the process that is used to reduce whole sheets of drawings to miniature size for ease of storage?

Answers to Questions

1. multilith offset
2. diazo process
3. the stenofax mimeograph machine
4. thermography
5. photocopying

Glossary

Aggregate Inert material mixed with cement and water to produce concrete.

Air entrained concrete Concrete in which a small amount of air is trapped by addition of a special material to produce greater durability.

Alkyd paint Paint with a binder consisting of synthetic resin producing a quick drying, tough paint surface.

American bond Brickwork pattern consisting of five courses of stretchers followed by one bonding course of headers.

Ampacity Current-carrying capacity expressed in amperes.

Anti-siphon trap Trap in a drainage system designed to preserve a water seal by defeating siphonage.

Apron Piece of horizontal wood trim under the sill of the interior casing of a window.

Appliance Utilization equipment, generally equipment other than industrial, normally built in standardized sizes or types and installed or connected as a unit to perform one or more functions, such as clothes washing, air-conditioning, food preparation, etc.

Approved Acceptable to the authority enforcing the Code.

Areaway Open space below the ground level immediately outside of a building; enclosed by substantial walls.

Arrester Wire screen secured to the top of an incinerator to confine sparks and other products of burning.

Ashlar Squared and dressed stones used for facing a masonry wall. Short upright wood pieces extending from the attic floor to the rafters forming a dwarf wall.

Backfill Loose earth placed outside foundation walls for filling and grading.

Balloon framing System of small house framing: two by fours extending two stories with inch by quarter ledger strips notched into the studs to support the second story floor beams.

Balustrade Protective or decorative railing.

Bargeborad Ornamented board covering the roof boards and projecting over the slope of the roof.

Batten Narrow wood strips used to cover joints. Also wood strip used to secure adjoining boards.

Batter Slope of the exposed face of a retaining wall.

Bead Narrow projecting molding with a rounded surface; or in plastering, a metal strip imbedded in plaster at the projecting corner of a wall.

Bearing plate   Steel plate placed under one end of a beam or truss for load distribution.

Bearing Wall   Wall supporting of a load other than its own weight.

Bed   Place or material in which stone or brick is laid; horizontal surface of positioned stone; lower surface of brick, stone, or tile.

Bell and spigot   Pipe joint formed with sections of cast iron pipe with a wide opening (bell) at one end and a narrow end (spigot) at the other, fitted by caulking with oakum and lead.

Belt course   Decorative horizontal band of masonry.

Bench mark   Point of reference from which measurements are made.

Billet   Heavy steel plate placed on concrete providing support for a column.

Blind nailing   Nailing together two wood members so that the nailheads do not show on the face of the work.

Block bridging   Solid wood members nailed between joists to stiffen floor.

Borings   Taking sample cylinders at varying depths of subsurface material at a proposed building site to determine the character of bearing material.

Braced frame   System of wood house framing using posts, girts, and braces.

Branch circuit   That portion of a wiring system extending beyond the final overcurrent device protecting the circuit.

Bridging   System of bracking between floor beams to distribute floor load.

Bridle iron   Steel hangers or stirrups shaped to secure wood beams together, such as headers to trimmers or tail beams to headers.

Brown coat   In three-coat plastering, the second coat of plaster troweled over the scratch coat; provides base for the white or finish coat.

Buck   Rough wood door frame placed in a wall or partition to which the door moldings are attached; completely fabricated steel door frame set in a wall or partition to receive the door.

Buttress   Projecting structure built against a wall to give it greater strength.

Caisson   Sunken panel in a ceiling, contributing to a pattern.

Calking compound   Mastic used to seal joints of wall openings against water.

Cantilever   Projecting beam or member supported at only one end.

Cant strip   Beveled strip placed in the angle between the roof and an abutting wall to avoid a sharp bend in the roofing material; strip placed under the lowest row of tiles on a roof to give it the same slope as the rows above it.

Casement window   Window sash opening on hinges secured to the side of the window opening.

Cavity wall   Wall built of solid masonry units arranged to provide air space within the wall.

Centering   Temporary wood framing supporting concrete forms.

Chamfer   Bevel edge surface area produced by cutting away external angle formed by 2 faces of stone or lumber.

Chase   Recess in inner face of masonry wall providing space for pipes and/or ducts.

Circuit breaker   A device designed to open and close a circuit by nonautomatic means and to open the circuit automatically on a predetermined overload of current, without injury to itself when properly applied within its rating.

Clerestory   Part of roof extending above main roof with windows.

Collar beam   Horizontal tie beam connecting rafters considerably above the plate.

Column   Vertical load carrying member of a structural frame.

Common or American bond   Brick laid in a pattern consisting of five courses of stretchers followed by one bonding course of headers.

Composite piles   A pile consisting of two different types of construction material in successive sections.

Contour line   On a land map denoting elevations, a line connecting points with the same elevation.

Coping   The highest course of a masonry wall bedded on the parapet wall.

Corbelling   Projecting courses of brick stepped out from the face of the wall forming a bracket for the wall above.

Cornerite   Strip of metal lath fitted into a corner to prevent cracking of the plaster.

Counter-flashing   Sheet metal strip in the form of an inverted L built into a wall to overlap the flashing and make the roof water-tight.

Cramp   Iron rod with ends bent to right angle; used to hold blocks of stone together.

Crawl space   Shallow space between the first tier of beams and the ground (no basement).

Cricket   Small false roof to throw off water from behind an obstacle.

Current   The rate of transfer of electricity.

Curtain wall   Non-bearing wall built between piers or columns for the enclosure of the structure; not supported at each story.

Dado   Decorative moulding on lower interior wall.

Dead man   Reinforced concrete anchor set in earth, tied to the retaining wall for stability.

Deflection   Deviation of the central axis of a beam from normal when beam is loaded.

Dentil   Repeated square, toothlike blocks, ornamental.

Dormer window   Extension from a sloped roof with a vertical window.

Double hung window   Window consisting of two sashes sliding vertically in adjoining grooves.

Double strength glass   One eighth inch thick sheet glass (single strength glass is one tenth of an inch thick).

Drip   Projecting horizontal course sloped outward to throw water away from building.

Dry wall   Interior wall construction consisting of plaster boards, wood paneling, or plywood nailed directly to the studs without application of plaster.

Dwarf partition   Partition which ends short of the ceiling.

Efflorescence   Blemish on the outside of brick walls consisting of a white surface crust formed from the crystallizing of soluble salts in the mortar.

Elevation   Drawing showing the projection of a building on a vertical plane.

End bearing pile   Pile acting like a column; the point has a solid bearing in rock or other dense material.

English bond   Pattern in brickwork consisting of alternate courses of headers and stretchers.

Entasis   Slight convexity of a column designed to make it pleasing to the eye.

Expansion bolt   Bolt with a casing arranged to wedge the bolt into a masonry wall to provide an anchorage.

Expansion joint   Joint between two adjoining concrete members arranged to permit expansion and contraction with changes in temperature.

Facade   Main front of a building.

Face brick   Brick selected for appearance in an exposed wall.

Factor of safety   Ratio of ultimate strength of material to maximum permissible stress in use.

Finish plaster   Final or white coat of plaster.

Fire brick   Brick made to withstand high temperatures for lining chimneys, incinerators, and similar structures.

Fire rated doors   Doors designed to resist standard fire tests and labeled for identification.

Fire stop   Incombustible filler material used to block interior draft spaces.

Fireproof wood   Chemically treated wood; fire resistive, used where incombustible materials are required.

Flashing   Strips of sheet metal bent into an angle between the roof and wall to make a water-tight joint.

Flat slab construction   Reinforced concrete floor construction of uniform thickness; eliminates the drops of beams and girders.

Flemish bond   Pattern of bonding in brickwork consisting of alternate headers and stretchers in the same course.

Flitch beam   Built-up beam consisting of a steel plate sandwiched between wood members and bolted.

Footing   Structural unit used to distribute loads to the bearing materials.

Frequency   The number of complete cycles an alternating electric current, sound wave, or vibrating object undergoes per second.

Friction pile   Pile with supporting capacity produced by friction with the soil in contact with the pile.

Frost line   Deepest level below grade to which frost penetrates in a geographic area.

Furring   Wood or metal strips nailed to walls or ceilings to provide a base for an even plastered surface; in walls, provides an air space between lath and wall to prevent condensation.

Gambrel roof   Roof with its slope broken by an obtuse angle.

Garden bond   Bond formed by inserting headers at wide intervals.

Girt   Heavy timber framed into corner posts as support for the building.

Government anchor   A v-shaped anchor, usually made of half-inch round bars to secure the steel beam to masonry.

Grade beam   Horizontal, reinforced concrete beam between two supporting piers at or below ground supporting a wall or structure.

Grillage   Framework of steel in a foundation designed to spread a concentrated load over a wider area; generally enclosed in concrete.

Groined ceiling   Arched ceiling consisting of two intersecting curved planes.

Ground   A conducting connection, whether intentional or accidental, between an electrical circuit or piece of equipment, and earth or some other conducting body serving in place of the earth.

Grounded   Connected to earth or to some conducting body that serves in place of the earth.

Grounds   Narrow strips of wood nailed to walls as guides to plastering and as a nailing base for interior trim.

Grout   Thin mortar, fluid enough to be poured into narrow spaces.

Gusset plate   Bracing steel plate to which pairs of angles are riveted forming a joint or bracket.

Half lap joint   Joint formed by cutting away half the thickness of each piece.

H-beam   Steel beam with wider flanges than an I-beam.

Header   Brick laid with an end exposed in the wall; wood beam set between two trimmers and carrying the tail beams.

Heading bond   Pattern of brick bonding formed with headers.

High tension bolts   High strength steel bolts tightened with calibrated wrenches to high tension; used as a substitute for conventional rivets in steel frame structures.

Hip jack rafter   Short rafter extending from the plate to the hip ridge.

Hip rafter   Rafter extending from the plate to the ridge forming the angle to a hip roof.

Hip roof   Roof with sloping sides and sloping end.

I-beam   Rolled steel beam or built-up beam of I section.

Incombustible material   Fire prevention material which will not ignite or actively support combustion in a surrounding temeprature of 1200°F during an exposure of 5 minutes; will not melt when the temperature of the material is maintained at 900°F for a period of at least 5 minutes.

Jack rafter   Short rafter used in hip or valley framing.

Jamb   Upright member forming the side of a door or window opening.

Joist and plank   Pieces of lumber with nominal dimensions of two to four inches in thickness by four inches and wider, of rectangular cross section and graded with respect to their strength in bending when loaded either on the narrow face as a joist or on the wide face as a plank.

Junior beam   Light weight structural steel sections rolled to a full I-beam shape.

Keyway   Groove formed in the top of a footing to anchor the foundation wall above; any groove formed in poured concrete to receive a succeeding pour.

King post   Central vertical tie in a truss.

Knee brace   Diagonal member placed across inside angle of framework to stiffen the frame.

Lally column   Compression member consisting of a steel pipe filled with concrete under pressure.

Laminated wood   Wood built up of piles or laminations that have been joined either with glue or with mechanical fasteners. Usually, the plies are too thick to be classified as veneer and the grain of all plies is parallel.

Lap joint   Joint between two wood members in which the same width and depth of the members is retained.

Leader   Vertical sheet metal pipe conducting rain from the roof gutter.

Ledger   In balloon framing, the board notched into the exterior studs supporting the 2nd story floor beams.

Let in braces   In wood house framing, the diagonal braces notched into studs.

Light   A pane of glass.

Lighting outlet   An outlet intended for the direct connection of a lampholder, lighting fixture, or pendant cord terminating in a lampholder.

Lintel   Horizontal steel member spanning an opening to support the load above.

Live load   All loads on structures other than dead loads; includes the weight of persons occupying the building and free standing material.

Lock seam   Joining of two sheets of metal consisting of a folded, pressed, and soldered joint.

Loft   Upper floor of a business building; wide floor area without partitions.

Mansard roof   Roof having on all sides two slopes, the lower slope being steeper than the upper one.

Mat foundation   Continuous reinforced concrete foundation constructed under entire building as a unit. (Also known as raft foundation or floating foundation.)

Membrane waterproofing   System of waterproofing masonry walls with layers of felt, canvas, or burlap and pitch.

Millwork   Finished wood products manufactured in millwork or planing mills such as window frames, windows and doors, stairways, interior trim, etc. Does not include flooring, siding, or ceiling.

Mullion   Vertical member forming a division between adjoining windows.

Muntin   Narrow bar separating window lights of a sash.

Nailing blocks   Wood members set on masonry to anchor other members to the masonry with nails or screws.

Newel post   Stairway post to which a handrail is secured.

Non-bearing wall   Wall which carries no load other than its own weight.

One way slab   Concrete slab with reinforcing steel rods providing a bearing on two opposite sides only.

Open web joist   Steel joists built up out of light steel shapes with an open latticed web.

Outlet   In the wiring system, a point at which current is taken to supply utilization equipment.

Parapet   Part of a masonry wall extending above the roof.

Pilaster   Flat square column attached to a wall and projecting about a fifth of its width from the face of the wall.

Piles   Long, slender members of wood, steel, or reinforced concrete driven into the ground to carry a vertical load.

Pintle   Iron member used at base of a wood post to anchor it into place; generally used with a cast iron cap over the post below.

Plate girder   Built up girder resembling an I beam with a web of steel plate and flanges of angle iron.

Platform framing   System of wood frame house construction using wood studs one story high finished with a platform consisting of the underflooring for the next story.

Plenum   Chamber or space forming a part of an air-conditioning system.

Plinth   Lowest member of a base or pedestal.

Post and girt   Wood framed buildings consisting of load bearing wood posts, widely spaced, connected with horizontal members called girts.

Power   The rate of doing work or expending energy.

Precast concrete   Concrete units (such as piles or vaults) cast off the construction site and set in place.

Prestressed concrete   System for utilizing fully the compressive strength of concrete by bonding it with highly stressed tensile steel.

Purlin   Horizontal members of roof that rest on roof trusses and support rafters.

Queen post   One of two vertical posts in a roof truss.

Quoin   Corner blocks of masonry. Stone or brick set at the corner of a building in blocks forming a decorative pattern.

Raceway   Any channel designed expressly for holding wire, cables, or bus bars and used solely for this purpose.

Raked joint   Joint formed in brickwork by raking out some of the mortar an even distance from the face of the wall.

Reflective insulation   Thin sheets of metal or foil on paper set in the exterior walls of a building to reflect radiant energy.

Relief valve   Safety device to permit the escape of steam or hot water subjected to excessive pressures or temperatures.

Return   The end railings of a fire-escape balcony.

Reveal   Space between window or door frame and the outside edge of the wall.

Ridge pole   Highest horizontal member of roof receiving upper ends of rafters.

Riser   Upright member of stair extending from tread to tread.

Roughing in   Installation of all concealed plumbing pipes; includes all plumbing work done before setting of fixtures or finishing.

Rowlock   Pattern of brickwork consisting of a course of brick laid on edge with ends exposed.

Running bond   Brick bond consisting entirely of stretchers.

Rusticated work   Squared stones with edges beveled or grooved to make the joints stand out.

Saddle   Short horizontal member set on top of a post to spread the load of the girder over it; piece of wood, stone, or metal placed under a door.

Sash balance   A device, dispensing with weights, pulleys, or cord, operated with a spring to counter-balance a double-hung window sash.

Scratch coat   First coat of plaster forming the key to the lath and the base for the second or brown coat.

Seat angle   Small steel angle riveted to one member to support the end of a beam or girder.

Separator   Sections of steel pipe forming spacers between I-beams bolted together serving as a structural unit.

Service   The conductors and equipment used for delivering energy from the electricity supply system to the wiring system of the premises served.

Service cable   The service conductors made up in the form of a cable.

Service conductors   The supply conductors that extend from the street main or transformers to the service equipment of the premises being supplied.

Service drop   The overhead service conductors from the last pole, or other aerial support, to and including the splices, if any, that connect to the service-entrance conductors at the building or other structure.

Service-entrance conductors, underground system   The service conductors between the terminals of the service equipment and the point of connection to the service lateral.

Service equipment   The necessary equipment, usually consisting of a circuit breaker, or switch and fuses, and their accessories, located near the point of entrance of supply conductors to a building and intended to constitute the main control and means of cutoff for the supply to that building.

Service lateral   The underground service conductors between the street main, including any risers at a pole or other structure or from transformers, and the first point of connection to the service-entrance conductors in a terminal box, meter, or other enclosure with adequate space, inside or outside the building wall. Where there is no terminal box, meter, or other enclosure with adequate space, the point of connection shall be considered to be the point of entrance of the service conductors into the building.

Sheathing   First covering of boards or paneling nailed to the outside of wood studs of frame building.

Sheave beams   Steel beams forming the overhead supports for an elevator.

Shim   Thin pieces of material used to bring members to an even or level bearing.

Shiplap   Wood boards cut with a rabbet or groove at opposite edges to provide an overlapping joint.

Shoved joint   Mortar joint produced by laying brick in a thick bed of mortar and forming a vertical joint of mortar by pushing the brick against the brick already laid in the same course.

Siding   Finishing material nailed to the sheathing of wood frame buildings forming the exposed surface.

Sill   Horizontal timber forming the lowest member of a wood frame house; lowest member of a window frame.

Skeleton construction   Buildings constructed of steel frame with the enclosure walls supported at each story.

Sleeper   Wood strips imbedded in concrete to provide a nailing base for the underflooring.

Soffit   Underside of a stair, arch, or cornice.

Soleplate   Horizontal bottom member of wood stud partition.

Solid bridging   Braces between floor joists consisting of short pieces with the same cross section as the joists.

Soil stack   Vertical cast iron pipe conveying sewage from branch waste pipes to house sewer.

Soldier course   Course of brick consisting of brick set on end with the narrow side exposed.

Spandrel   In steel skeleton construction, the outside wall from the top of a window to the sill of the window above.

Spread footing   Footing designed for wider bearing on weak soils; often with reinforcing steel and of shallow depth in proportion to width.

Stack   Any vertical line of soil, waste, or vent piping.

Stirrup   Metal strap of U form supporting one end of a wood beam.

Stretcher   Brick laid with its length parallel to the wall and side exposed.

Stringer   Members supporting the treads and risers of a stair.

Strut   A compression member other than a column or pedestal.

Studs   Vertically set skeleton members of a partition or wall to which lath is nailed.

Switch, general-use   A switch intended for use in general distribution and branch circuits. It is rated in amperes and is capable of interrupting its rated voltage.

Switch, general-use snap   A form of general-use switch so constructed that it can be installed in flush device boxes or on outlet covers, or otherwise used in conjunction with wiring systems recognized by the Code.

Switch, AC general-use snap   A form of general-use snap switch suitable only for use on alternating-current circuits and for controlling the following:
1. Resistive and inductive loads (including electric discharge lamps) not exceeding the ampere rating at the voltage involved.
2. Tungsten-filament lamp loads not exceeding the ampere rating at 120 volts.
3. Motor loads not exceeding 80% of the ampere rating of the switches at the rated voltage.

Switch, AC-DC general-use snap   A form of general-use snap switch suitable for use on either direct- or alternating-current circuits and for controlling the following:
1. Resistive loads not exceeding the ampere rating at the voltage involved.
2. Inductive loads not exceeding one-half the ampere rating at the voltage involved, except that switches having a marked horsepower rating are suitable for controlling motors not exceeding the horsepower rating of the switch at the voltage involved.
3. Tungsten-filament lamp loads not exceeding the ampere rating at 125 volts, when marked with the letter T.

Switchboard   A large single panel, frame, or assembly of panels, having switches, overcurrent, and other protective devices, buses, and usually instruments, mounted on the face or back or both. Switchboards are generally accessible from the rear as well as from the front and are not intended to be installed in cabinets.

Terne-plate   Sheet iron coated with an alloy of four parts of lead to one part of tin.

Terrazzo   Stone floor consisting of marble chips laid in concrete patterned with brass strips.

Ties   A tension member.

Transformer A device used to transfer energy from one circuit to another. It is composed of two or more coils linked by magnetic lines of force.

Trap U-shaped bend in drain pipe providing space for water seal.

Travertine Porous marble building stone.

Trimmer Beam framing an opening in a wood joist floor supporting the header beam.

Trusses Framed structural pieces consisting of triangles in a single plane for supporting loads over spans.

Vermiculite Lightweight inert material made of steam exploded mica used as an aggregate in plaster.

Volt The practical unit of voltage or electromotive force. One volt sends a current of one ampere through a resistance of one ohm.

Voltage (of a circuit) The greatest root-mean-square (effective difference of potential) between any two conductors of the circuit concerned.

Voltage to ground In grounded circuits, the voltage between the given conductor and that point or conductor of the circuit which is grounded; in ungrounded circuits, the greatest voltage between the given conductor and any other conductor of the circuit.

Watt The practical unit of electrical power.

Web Central portion of an I-beam.

Withe Partition between flues of chimney.

Index

Other Practical References

■ National Construction Estimator

Current building costs for residential, commercial, and industrial construction. Estimated prices for every common building material. Manhours, recommended crew, and labor cost for installation. Includes Estimate Writer, an electronic version of the book on computer disk, with a stand-alone estimating program ---- free on 5¼" high density (1.2Mb) disk. The National Construction Estimator and Estimate Writer on 1.2Mb disk cost $31.50. (Add $10 if you want Estimate Writer on 5¼" double density 360K disks or 3½" 720K disks.) **592 pages, 8½ x 11, $31.50. Revised annually**

■ Estimating Home Building Costs

Estimate every phase of residential construction from site costs to the profit margin you include in your bid. Shows how to keep track of manhours and make accurate labor cost estimates for footings, foundations, framing and sheathing finishes, electrical, plumbing, and more. Provides and explains sample cost estimate worksheets with complete instructions for each job phase. **320 pages, 5½ x 8½, $17.00**

■ Video: Stair Framing

Shows how to use a calculator to figure the rise and run of each step, the height of each riser, the number of treads, and the tread depths. Then watch how to take these measurements to construct an actual set of stairs. You'll see how to mark and cut your carriages, treads, and risers, and install a stairway that fits your calculations for the perfect set of stairs. **60 minutes, VHS, $24.75**

■ Planning Drain, Waste & Vent Systems

How to design plumbing systems in residential, commercial, and industrial buildings. Covers designing systems that meet code requirements for homes, commercial buildings, private sewage disposal systems, and even mobile home parks. Includes relevant code sections and many illustrations to guide you through what the code requires in designing drainage, waste, and vent systems. **192 pages, 8½ x 11, $19.25**

■ Electrical Blueprint Reading Revised

Shows how to read and interpret electrical drawings, wiring diagrams, and specifications for constructing electrical systems. Shows how a typical lighting and power layout would appear on a plan, and explains what to do to execute the plan. Describes how to use a panelboard or heating schedule, and includes typical electrical specifications. **208 pages, 8½ x 11, $18.00**

■ Contractor's Guide to the Building Code Revised

This completely revised edition explains in plain English exactly what the Uniform Building Code requires. Based on the most recent code, it covers many changes made since then. Also covers the Uniform Mechanical Code and the Uniform Plumbing Code. Shows how to design and construct residential and light commercial buildings that'll pass inspection the first time. Suggests how to work with an inspector to minimize construction costs, what common building shortcuts are likely to be cited, and where exceptions are granted. **544 pages, 5½ x 8½, $28.00**

■ Residential Electrical Design

Shows how to draw up an electrical plan from blueprints, including the service entrance; grounding; lighting requirements for kitchen, bedroom and bath; and how to lay them out. Explains how to plan electrical heating systems and what equipment you'll need, how to plan outdoor lighting, and much more. If you're a builder who sometimes has to plan an electrical system, you should have this book. **194 pages, 8½ x 11, $11.50**

■ Construction Estimating Reference Data

Provides the 300 most useful manhour tables for practically every item of construction. Labor requirements are listed for sitework, concrete work, masonry, steel, carpentry, thermal and moisture protection, door and windows, finishes, mechanical and electrical. Each section details the work being estimated and gives appropriate crew size and equipment needed. This new edition contains *DataEst*, a computer estimating program on a high density disk. This fast, powerful program and complete instructions are yours free when you buy the book. **432 pages, 11 x 8½, $39.50**

■ Wood-Frame House Construction

Step-by-step construction details, from the layout of the outer walls, excavation and formwork, to finish carpentry and painting. Contains all new, clear illustrations and explanations updated for construction in the '90s. Everything you need to know about framing, roofing, siding, interior finishings, floor covering and stairs -- your complete book of wood-frame homebuilding. **320 pages, 8½ x 11, $19.75. Revised edition**

■ Carpentry Layout

Explains the easy way to figure: cuts for stair carriages, treads and risers; lengths for common, hip, and jack rafters; spacing for joists, studs, rafters, and pickets; layout for rake and bearing walls. Shows how to set foundation corner stakes, even for a complex home on a hillside. Practical examples on how to use a hand-held calculator as a powerful layout tool. **240 pages, 5½ x 8½, $16.25**